Lignin - Chemistry, Structure, and Application

Edited by Arpit Sand and Jaya Tuteja

Published in London, United Kingdom

Lignin - Chemistry, Structure, and Application
http://dx.doi.org/10.5772/intechopen.102263
Edited by Arpit Sand and Jaya Tuteja

Contributors
Jost Ruwoldt, Tawaf Ali Shah, Li Zhiyu, Li Zhihe, Zhang Andong, Zhenglun Li, Ana Lourenço, Jorge Gominho, Mehran Rahimlou, Jalal Hejazi, Sherif Mehanny, Mohamed H. El-Moayed, Mahmoud Farag, Justin Kühn, Sea-Hyun Lee, Arpit Sand, Jaya Tuteja

First published in London, United Kingdom, 2023 by IntechOpen
IntechOpen is the global imprint of INTECHOPEN LIMITED, registered in England and Wales, registration number: 11086078, 5 Princes Gate Court, London, SW7 2QJ, United Kingdom

British Library Cataloguing-in-Publication Data
A catalogue record for this book is available from the British Library

Additional hard and PDF copies can be obtained from orders@intechopen.com

Lignin - Chemistry, Structure, and Application
Edited by Arpit Sand and Jaya Tuteja
p. cm.
Print ISBN 978-1-83968-545-3
Online ISBN 978-1-83968-546-0
eBook (PDF) ISBN 978-1-83968-861-4

Meet the editors

Dr. Arpit Sand is currently an associate professor in the Department of Chemistry, at Manav Rachna University, Faridabad, India. He received his BSc in science and MSc in chemistry from the University of Allahabad, India, in 2004 and 2006, respectively, followed by a Ph.D. in chemistry from the same institution in 2010. Dr. Sand's previous roles include assistant professor (guest faculty) in the Department of Chemistry, University of Allahabad; research associate at the National Physical Laboratory (NPL) New Delhi; postdoctoral fellow at Gwangju Institute of Science & Technology (GIST), Korea and at Karolinska Institutet, Sweden; junior researcher at Brno University of Technology, Czech Republic; and researcher at Soongsil University, South Korea. Dr. Sand's research interests include polymer synthesis using different polymerization techniques. He has made significant contributions to the modification and characterization of graft copolymers and films decorated with chalcogenide quantum dots to tune the energy band gap for solar energy harvesting applications. Dr. Sand has authored more than 24 international research articles and review articles in SCI journals. He is a member of the editorial boards of *Polymer Synthesis Journal and Journal of Polymer Science* and a reviewer for international journals including *Carbohydrate Polymers, International Journal of Biological Macromolecules, and Fibers and Polymers.* He is also a life member of the Indian Science Congress and Green Chemistry Network Center.

Dr. Jaya Tuteja has been an assistant professor in the Department of Chemistry at Manav Rachna University since 2018. She completed her Ph.D. in materials science in 2015 at the Japan Advanced Institute of Science and Technology (JAIST), where she obtained an Excellent Doctorate Student award. Prior to her current appointment, she worked in the chromatography industry as a Technical and Application Manager for YMC India Ltd. Her research interests include the development, characterization and application of heterogeneous catalysts for effective conversion of biomass to value-added products; nanoparticle synthesis and nanoparticle-based heterogeneous catalysts; bimetallic nanoparticles; carbohydrate chemistry; and analytical techniques for compound identification. She has published one book chapter and five research papers in reputed international journals with a cumulative impact factor of >25 and has two patents filed in her name. One of her papers was among the 25 Most Accessed Articles from *ChemSusChem* journal.

Contents

Preface

Lignin is the second-most abundant renewable polymer found in the cell walls of plants. Composed of various alkyl-aromatic polymer units, lignin has interesting physical and chemical properties making it suitable for application in various chemical industries. Its sole disadvantage is its rigidity. At present, the foremost use of lignin is combustion to generate power and heat, but in the last decade, catalysis has improved the efficiency of lignin processing for materials applications. Depolymerization is the most important process used in the preparation of lignin as a natural polymer. A great deal of research has been carried out to find efficient lignin depolymerization processes. Despite extensive development work, lignin remains underutilized in biorefineries.

This book describes the detailed structure of lignin, its chemical properties and physical significance, and its chemical and materials applications. It provides a brief history of lignin and materials derived from it by modification and describes recent research in lignin applications for structural materials, raw polymers, adhesives, plastics, and many more. It also highlights the challenges associated with maximizing output, with potential solutions.

Arpit Sand and Jaya Tuteja
School of Sciences,
Manav Rachna University,
Faridabad, Haryana, India

Section 1

Introduction to Lignin

Chapter 1

Introductory Chapter: Introduction to Structure Properties and Application of Lignin

Jaya Tuteja and Arpit Sand

1. Introduction

Here, the polymeric properties of naturally existing lignin and its potential in the field of many chemical industries have been focused. The structure of lignin is considered as the cross-linking of various polyphenols containing methoxy groups and aldehyde functionality at terminal position. Lignin is the second most naturally existing polymer with full of aromatic functionality. Despite being its abundance in nature lignin, utilization is not seen in relevant areas such as polymers, adhesives, and rubber industries.

With exponential growth of population and their energy demands, there is a strong urge to incline our dependency on sustainable approaches [1]. The recent researches are focusing on the development of methods which are dependent on sustainable or renewable sources. Lignin is a nonedible polymer with carbon-neutral concept and is found in plenty of quantities from agriculture and forestry residues [2]. For the low utilization of these lignin resources and considering its potential in various chemical industries, the researchers are continuously working to develop methods for isolation of lignin (high-quality technical) for further transformation to various value-added chemicals [2].

2. Structure of lignin

As known from centuries, that lignin is found in the cell wall of the plants and is the major component to provide rigidity to cell walls. The cell wall is basically constituted of cellulose, hemicellulose, and lignin part. The utilization of cellulose and hemicellulose can be seen in various forms in literature, but lignin owing to its complex polymeric structure not being utilized much. As mentioned above, lignin is composed of various aromatic units along with methoxy and aldehyde groups as their functionality. The typical composition of these monomer units is different in different plant species dependent on their growth, their environmental condition, type of biomass, etc. [3, 4]. Despite seeing their variation in composition, it is generally considered that every cell wall on an average contains 15–25% of lignin along with 30–40% of cellulose and 15–30% hemicellulose, and few % of other components like starch, pectin, protein, carbohydrate, minerals, etc., exist in the cell wall [5–7]. Most of the natural polymers are made up of single monomer; on contrary, lignin is composed of 3-D copolymer interconnected via ether linkages of phenylpropanoid units (**Figure 1**).

IntechOpen

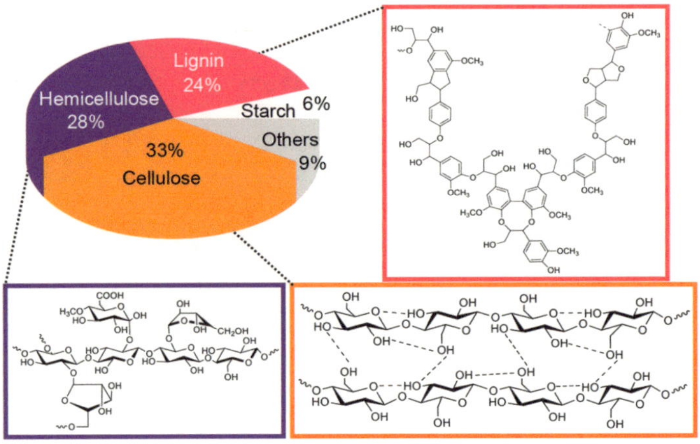

Figure 1.
Structure of lignocellulosic biomass.

3. Characterization of lignin

It is very important to know the composition and structure well before planning to use them as precursor for any of the applications. Since lignin can be recovered from various plant resources and can be separated well by numerous methods, identification of its unique characteristics is very important [8, 9]. The functional group analysis in lignin is conducted by using FT-IR spectroscopy and Raman and fluorescence spectroscopy [10]. The stretching frequencies for ether bond, methoxy, and hydroxyl groups can be easily identified using FT-IR spectroscopy [11]. Macromolecular structure of lignin is determined by electron microscopy such as SEM-EDX (scanning electron microscope with energy dispersive X-ray spectroscopy) and TEM (transmission electron microscopy). UV-visible spectroscopy shows good absorption band owing to its aromatic nature. The shape of UV-visible graph is very sensitive to its particular type, pH, and solvent used [12].

Many structural aspects of lignin regarding its composition and reactivity can be assessed by using advanced NMR techniques like 2D-NMR and 2D HSQC-NMR. The basic and rough estimation of lignin structure can be done by [13] C-NMR and [1] H-NMR, but the spectra obtained is much overloaded and complex. For better resolution, hyphenated NMR techniques gives best results [13–15]. Gel polymer chromatography (GPC) is a type of size exclusion chromatography which is utilized in measuring the molecular weight of lignin polymers. Various aqueous and nonaqueous solvents can be used based on hydrophilic and hydrophobic nature of lignin molecule [16, 17].

4. Application of lignin

Target application is the only thing which defines which lignin molecule (of which particular characteristics) is required. As a general rule, we can state that low-quality lignin is used for the production of lignin-based polymers or plastics, whereas high-quality lignin is used for their transformation to value-added products of biomedical applications [18–20]. Lignin is used as food packaging material because of its antioxidant properties due to aromatic ring with hydroxyl and methoxy groups functionalities present in it. It has been noticed from the literature that lignin-based food packaging helps to protect the food against UV radiation [21]. High potential of lignin is also observed in biofoaming. Foaming in polymers is introduced by blowing air or chemicals between the polymer matrixes. Here, the lignin has found its importance due to polyhydroxyl groups present in it, as it can substitute natural polyols in polyurethane [22]. Resins are the another important sector where lignin has been found replacing phenol in phenol-formaldehyde resins [23].

Considering the tremendous properties of lignin, it has been an interest of researchers to find out its biomedical and pharma applications in form of hydrogels, aerogels, and sensors. Lignin-based hydrogels have been noticed to show good results in tissue engineering and drug delivery applications owing to their porous nature and high swelling capacity in aqueous media [24]. A review article published in 2000 has shown the application of lignin in biosensing [25]. These lignin-based sensors are designed to measure the biological species quantitatively.

Apart from above-mentioned applications, there are still endless applications of lignin molecule. It is really difficult to task to enlist all the applications here. Few such applications are absorption of heavy metals [26], flocullants [27], fibers [28], batteries [29], and much more.

Author details

Jaya Tuteja* and Arpit Sand
School of Sciences, Manav Rachna University, Faridabad, Haryana, India

*Address all correspondence to: jayatuteja89@gmail.com

References

[1] Delgado N, Ysambertt F, Chavez G, Bravo B, Garcıa DE, Santos J. Valorization of Kraft Lignin of Different Molecular Weights as Surfactant Agent for the Oil Industry. Waste and Biomass Valorization. 2019;**10**:3383-3395

[2] Melro E, Filipe A, Sousa D, Medronho B, Romano A. Revisiting lignin: A tour through its structural features, characterization methods and applications. New Journal of Chemistry. 2021;**45**:6986-7013

[3] Agbor VB, Cicek N, Sparling R, Berlin A, Levin DB. Biomass pretreatment: Fundamentals toward application. Biotechnology Advances. 2011;**29**:675-685

[4] Sawatdeenarunat C, Surendra KC, Takara D, Oechsner H, Khanal SK. Anaerobic digestion of lignocellulosic biomass: Challenges and opportunities. Bioresource Technology. 2015;**178**: 178-186

[5] Alonso DM, Wettstein SG, Dumesic JA. Bimetallic catalysts for upgrading of biomass to fuels and chemicals. Chemical Society Reviews. 2012;**41**:8075-8098

[6] Isikgor FH, Becer CR. Lignocellulosic biomass: a sustainable platform for the production of bio-based chemicals and polymers. Polymer Chemistry. 2015;**6**:4497-4559

[7] Frigon JC, Guiot SR. Biomethane production from starch and lignocellulosic crops: a comparative review. Biofuels, Bioproduction Biorefining. 2010;**4**:447-458

[8] Stark NM, Yelle DJ, Agarwal UP. Lignin. Polymer Composites. 2016:49-66

[9] Allison Tolbert AJR, Akinosho H, Khunsupat R, Naskar AK. Characterization and analysis of the molecular weight of lignin for biorefining studies. Biofuels, Bioproducts & Biorefining. 2014;**8**:836-856

[10] You T, Xu F. Applications of Molecular Spectroscopy to Current Research in the Chemical and Biological Sciences. 2016;**2**:235-260

[11] Kallavus U, Karner K, Elomaa M. Rapid semi-quantitative determination of aspen lignin in lignocellulosic products. Proceedings of the Estonian Academy of Sciences. 2015;**64**:105

[12] Skulcova A, Majova V, Kohutova M, Grosik M, Sima J, Jablonsky M. UV/Vis Spectrometry as a Quantification Tool for Lignin Solubilized in Deep Eutectic Solvents. BioResources. 2017;**12**:6713-6722

[13] Wen J-L, Yang S, Sun S-L, Yuan T-Q, Sun R-C. Lignin Structures Analytical Applied Biomaterial Ecological Significance. 2014:92-109

[14] Wen JL, Sun SL, Xue BL, Sun RC. Recent Advances in Characterization of Lignin Polymer by Solution-State Nuclear Magnetic Resonance (NMR) Methodology. Materials. 2013;**6**:359-391

[15] Chu S, Subrahmanyam AV, Huber GW. The pyrolysis chemistry of a β-O-4 type oligomeric lignin model compound Green Chemistry. 2013;**15**:125-136

[16] Sjostrom E, Alen R. Analytical Methods in Wood Chemistry, Pulping, and Papermaking. 1998

[17] Chen F, Li J. Aqueous Gel Permeation Chromatographic Methods for Technical

Lignins. Journal of Wood Chemistry and Technology. 2000;**20**:265-276

[18] Saito T, Brown RH, Hunt MA, Pickel DL, Pickel JM, Messman JM, et al. Turning renewable resources into value-added polymer: development of lignin-based thermoplastic. Green Chemistry. 2012;**14**:3295-3303

[19] Iravani S. Nanoparticles and their Biomedical Applications. Springer; 2020. pp. 217-224

[20] Witzler M, Alzagameem A, Bergs M, El Khaldi-Hansen B, Klein SE, Hielscher D, et al. Lignin-Derived Biomaterials for Drug Release and Tissue Engineering. Molecules. 2018;**1885**:23

[21] Shankar S, Reddy JP, Rhim JW. Effect of lignin on water vapor barrier, mechanical, and structural properties of agar/lignin composite films. International Journal of Biological Macromolecules. 2015;**81**:267-273

[22] Obaid N, Kortschot MT, Sain M. Lignin in Polymer Composites. Elsevier Inc; 2015. pp. 217-232

[23] Kai D, Tan MJ, Chee PL, Chua YK, Yap YL, Loh XJ. Towards lignin-based functional materials in a sustainable world. Green Chemistry. 2016;**18**:1175-1200

[24] Espinoza-Acosta JL, Torres-Chavez PI, Ramırez-Wong B, Lopez-Saiz CM, Montano-Leyva B. Antioxidant, antimicrobial, and antimutagenic properties of technical lignins and their applications. BioResources. 2016;**11**:5452-5481

[25] Moreno A, Sipponen MH. Lignin-based smart materials: a roadmap to processing and synthesis for current and future applications. Materials Horizons. 2020;**7**:2237-2257

[26] Maldhure AV, Ekhe JD. Effect of modifications of lignin on thermal, structural, and mechanical properties of polypropylene/modified lignin blends. Journal of Thermoplastic Composite Materials. 2017;**30**:625-645

[27] Wang B, Wang S-F, Lam SS, Sonne C, Yuan T-Q, Song G-Y, et al. A review on production of lignin-based flocculants: Sustainable feedstock and low carbon footprint applications. Renewable and Sustainable Energy Reviews. 2020;**134**:110384

[28] Fang W, Yang S, Wang X-L, Yuan T-Q, Sun R-C. Manufacture and application of lignin-based carbon fibers (LCFs) and lignin-based carbon nanofibers (LCNFs). Green Chemistry. 2017;**19**:1794-1827

[29] Lima RB, Raza R, Qin H, Li J, Lindstrom ME, Zhu B. Direct ligninfuel cell for power generation. RSC Advances. 2013;**3**:5083-5089

Chapter 2

Composition and Role of Lignin in Biochemicals

Tawaf Ali Shah, Li Zhihe, Li Zhiyu and Zhang Andong

Abstract

The term lignin is derived from lignum, which means plant wood. Plant wood are mainly composed of extractives, hemicellulose, cellulose, and lignin. The lignin is a cross-linked polymer, made of three phenylpropanoid precursors, *p*-coumaryl, synapyl, and conniferyl alcohols. It is the most abundant polymer in plant world and act mechanically as a natural glue to bind hemicellulose and cellulose. Lignin is amorphous, soluble in alkali, condenses with phenol and has high melting temperature. The function of lignin is to protect the carbohydrates of the biomass from degradation, thus provide stability. The chapter includes information on types of lignin, structure, isolation, degradation, and transformation in to market value chemicals. The application of lignin and lignin base monomers for synthesis of plastic, hydrogels, adhesives, chemicals, fuels and other value added materials at industrial scale.

Keywords: lignin, plant biomass, polymer, pretreatment, biorefinery

1. Introduction

In organic biomass, particularly in the plant biomass, lignin is one of the rich carbon matter. Lignin is part of plant cell wall, which provides a mechanical muscle to keep herbs, plants and large tree standing. The process of synthesis of lignin inside plant biomass is known as lignification. The lignin is a long amorphous polymer containing three major synapyl, *p*coumaryl alcohols and conniferyl phenylpropanoid monomers. Due to this aromatic polymeric nature, lignin supports the other carbohydrates components like cellulose and hemicellulose in lignocellulosic plant biomass. So its acts like a binder for the carbohydrates of plant biomass to make a strong cross link and bond between hemicellulose and cellulose [1].

An estimated quantity of 5 to 36 × 108 tons of lignin is reported annually in natural organic biomass [2]. The uppermost 26–45% of dry matter lignin part is naturally observed in softwood whereas 15–25% a lower amount of lignin in hardwoods biomass [3]. The value of lignin in grasses and straw is relatively low from the softwood range between 15 and 27%, and a small percentage in algae and mosses. As an evolutionary perception, the development of aromatic lignin structure in plant cell wall, is for fast water transportation inside the tracheid plant cell structure [2]. This is an important material of organic carbon polymer incorporation in plant biomass for possible growth in dry land. The lignin makes the biomass rigid and hard, possiblily can decrease the water loss in dry and drought condition.

1.1 Significance of biomasses

The word biomass means organic substance made of plants materials. This could be a forest wood, cereal crops, seaweed and leftover substrates from agriculture biomass. Other sources of biomass are human, animals, household and industrials waste. These biomasses contain chemical energy trapped directly from sun through photosynthesis. Photosynthesis converts carbon dioxide and water with the help of sun energy in to sugars units, that's make different carbohydrates in fruits, vegetables, and other crops plants. Biomasses for example straw, food like potato, cassava, corn starch, sorghum, wastewaters and other agro-industrial residues are a good source of sugars (glucose, xylose, arabinose, starch, sucrose, cellulosic, and reducing sugars) that can be recycle for multiple products at commercial level. Biomasses (softwood, hardwood, herbs, straw and grasess) are also a good source of organic carbohydrates materials. On the basis of high organic contents, these biomass are favourit food for most of the living species on the globe. Beside, food purpose, the left over residue and waste biomass can also contribute about 7–10% to the global energy stream. These agriculture biomasses, softwood, hardwood, herbs, straw and grasess contained high carbon and water content, thus considered a significant input carbon for production of renewable energy like biodiesel, bioethanol, biogas, biohydrogen. Currently, 8–15% of the overall energy stream is reported in industrials zones of developed countries, while a low amount in developing countries as well [4, 5]. Beside, renewable energy, most of the countries practiced to burn these biomass either in the field or for cooking in stoves, thus a massive amounts of carbohydrates biomass containing useful carbon (lignin, cellulose and hemicellulose) is destroyed on low price [6]. Food waste and waste lignocellulosic biomass are of potential interest, which includes utilization for bio-methane, ethanol, butanol, biomaterials, and biobased products [7, 8]. Food waste is easily digestible in fermentation process for biofuel production [9], however, lignocellulosic biomass like grasses, waste paper, wood residues, and crop residues are diffficult to convert direct into fuel. Because the lignocellulosic biomass is comprised of mainly three basic molecules, 20–40% lignin, and 60–70% hemicellulose, cellulose and other extractives in collectlivelly [10]. The composition of carbohydrates varies depending on the growth and nature of the biomass. There is a big barrier in utlization of lignocellulosic biomass containing high lignin content, due to strong crosslinking of ether and ester bonds around polysaccharides (cellulose and hemicellulose) which confers hydrolytic stability, resist degradation and structural recalcitrance [5]. Therefore, to utilize these lignin containing biomasses, pretreatment process are necessory to make them digestible in the fermentation process and synthesis of value added materials.

Energy is the eventual source need for each operation through the globe. More than 25% of the World population is suffering the lack of energy for cooking, lighting, power generation and transportation fuel. The consumption of the energy is higher than the production percentage [11]. The demand for extra energy production builds an energy crisis throughout the world. The details reports released by the European Environment Agency and International Energy Agency (IEA), Energy Information Administration (EIA), oil usage and cost is mounting constantly. The published reports indicated some problems and hurdles in future for energy production and emphasized on potential alternate resolutions. The estimated amount of energy used for heat and power from oil 31.0%, natural gas 21.5%, nuclear 5%, coal 29.0%, hydro 3%, biofuels and waste 10.0%, and 'other' 4%. The electricity generation estimated data was natural gas 22%, oil 31.1%, coal 28.9%, nuclear 12%, hydro 13%, biofuels and waste 10.2%, and 'other' 6% as shown in the (**Figure 1**). These values specify that

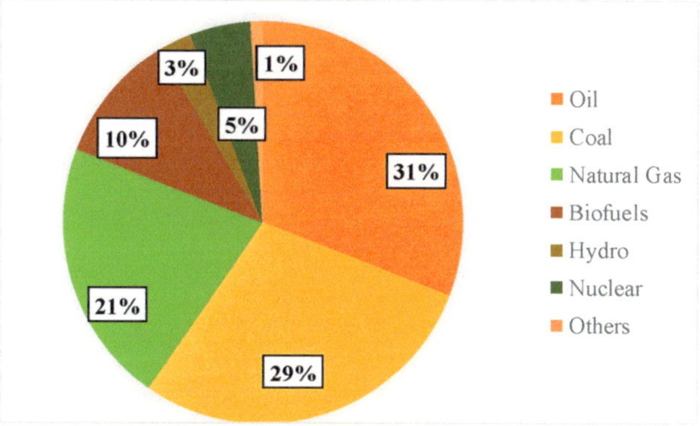

Figure 1.
The results shown here is replecting the energy production percentage around the globe with less ttention for biofuels from waste biomass.

generation of energy from alterantive sources like biofuels and bioenergy are receiving more attention instead of fossil fuels [12, 13]. Consumption of waste biomasses (agribiobiomass and municipal solid waste) for biofuels could meet a rational demand of energy percentage in the future.

To solve the problems of energy crises, production of renewable energy from waste biomass sources is getting more interest instead of fuel generation from fossil sources. Waste biomass especially lignocellulosic biomass residue along other biomassess is a good waste reserve that could be reuse for important products [14]. So, utilizing the waste biomass can produce clean bioenergy and has a benefit of to efficiently manage solid waste to reduce environmental pollution. The agriculture waste and waste residue which is about 60–70% around the world can be a cheap carbon material for biofuel and bioenergy production. The cellulose and hemicellulose are carbon source which can be transformed in to verities of fuels, which includes ethanol, butanol, bio-methane and biohydrogen [1, 15]. However, lignin is an aromatic component, cross link both cellulose and hemicellulose of the lignocellulosic biomass which makes it recalcitrant and harder, so decrease the yield of energy production [16]. Therefore, pretreatment process are carried out to remove the lignin, cellulosc and hemicellulsoe for the biofuel production. The separated lignin is also a good source of fine chemicals and industrially relevant materials discussed in details below.

For the improvement of effective anaerobic fermentation processs, the choice of waste biomass, nature and chemistry of biomass is mainly essential factors. The lignin percentage and quantity of water are the next important properties in picking substrates for the fermentation reaction. While the total organic carbon TOC is the most critical concern factor in the conversion of the waste carbon biomass [17, 18]. The efficiency of the microbial hydrolysis of the biomass in anaerobic fermentation reaction is linked to the biodegradability of biomass [19, 20].

2. Biomass components

The organic constituents in agricultural biomass have a recalcitrant structured cell wall which is called "lignocellulose" [21]. It is evident that the three components cellulose, lignin and hemicellulose are primery components as shown in the (**Figure 2** and **Table 1**). The highest composition is of cellulose in plant cell wall, whereas the

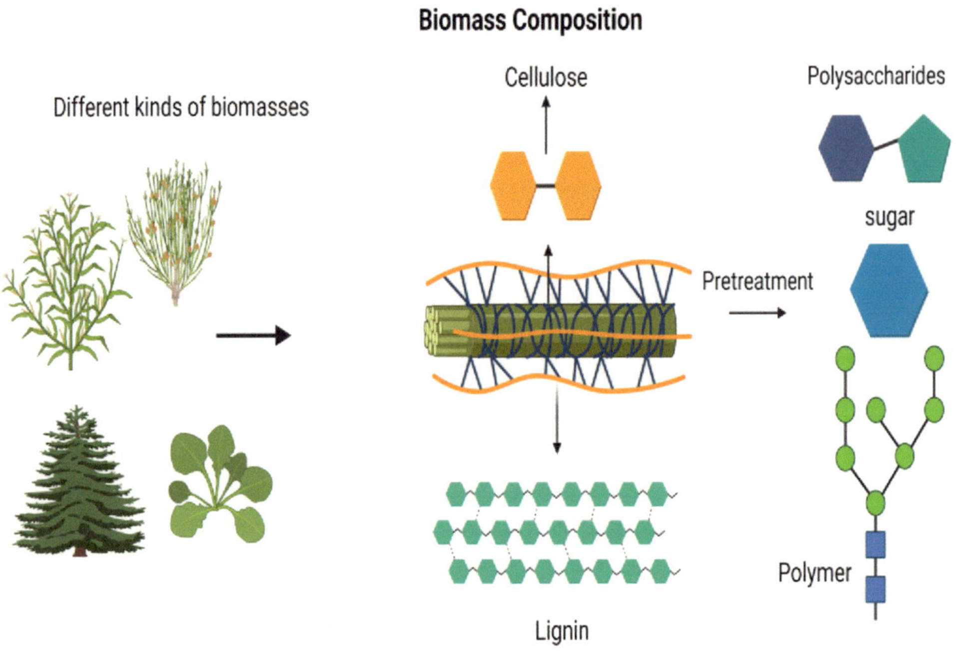

Figure 2.
An overview of the general components of plant biomasses.

Substrates	Cellulose	Hemicellulose	Lignin
Softwoods	45–50	25–35	25–35
Bamboo	41–49	24–28	24–26
Hardwoods	40–55	24–40	18–25
Paper	40–55	25–35	15–20
Corn cob	40–50	20–35	15–17
Peanut	40–45	15–17	20–30
Corn straw	38–45	26–34	17–20
Para grass	35–45	25–30	15–20
Corn stalk	34–36	26–27	16–21
Wheat straw	30–40	30–50	15–16
Sweet sorghum	27–38	21–25	11–17
Grasses	25–40	35–50	10–30
Switch grass	45	31.4	12
Rice straw	38	19	13
Barley straw	37	24	16
Sugarcane bagasse	33	23	5
Arundo donax	31	30	21
Lawn grass	30	43	3–5

Table 1.
Biomass main components and their percent value.

second copolymer is hemicellulose and the third one is aromatic protective polymer is of lignin. The components are discussed in the details separatly.

2.1 Cellulose

Cellulose is a homopolyme of long chain basic units of 4-O-β-D-glucopyranosyl-D-glucose and cellubiose. On hydrolysis of cellulosic polymer of most of biomasses is usually of glucose molecule. Chemically, the cellulose has a long chain of (C6H10O5) n units in the polymer. The number of units in cellulose polymer is about 700–1000 glucose units. These glucose units are bonded togather with β-1,4 glucosidic bond between the two connecting glucose in a long chain. The glucose units have strong hydrogen bonds in parrallel position to support further the strength between the molecules [22]. Cellulose has crystalline and non-crystalline structure, sorrunded by microfibrils to form fibers. Cellulose has the capacity to absorb about 7–15% water, and is not soluble in diluted acid and water under normal temperature. On increasing the temperature, acid concentration can increase solubility, whereas alkali solution separate, decrease degree of polymerization and swells the cellulose crystalline structure [10].

2.2 Hemicellulose

Hemicellulose is the second highest heterogeneous polymer comprises of various sugars molecules i.e. galactans, glucuronoxylan, glucomannan, and low amounts of other polysaccharides. The other sugars molecules are glucans, xylans, galactans, arabinans and mannans exist in the hemicellulose molecule. The glucans, galactans, arabinans and mannans are linked togather with β-1,4 linked backbone. In the hemicellulose polymer, xylan contains 1–4 linkages of xylopyranosyl units attached with α-4-O-methyl-D-glucuronopyranosyl branch chain of glucose connected to other xylose carbon sugar monomers [23]. Notably at ordinary temperature, the hemicellulose is also insoluble in water. It is soluble in acidic solution, and the solubility increase with concnetration of acid and temperature. Naturally, hemicellulose has various size, shape, low polymerization in structure and is not crystalline owing acetyl groups in the polymer. The hemicellulose make a strong backbon in plant biomass by crosslinkg with non-covalent bonds cellulose micro-fibril chains [23].

2.3 Lignin

Lignin is different than the hemicellulose and cellulose because it comprise of aromatic heterogeneous polymer phenyl-propane units. The long phenyl-propane units are attached togather by ether bonds. The main function of lignin is to strongly drag hemicellulose and cellulose in complex structure of lignin like a glue to seal each gap in plant biomass. Therefore, this lignin make it harder for the degradation. The concentration of lignin is depending on the size, growth, nature and properties of plants biomass [24]. It is soluble in varios solvents, dioxane, pyridine, dimethyl sulfoxide, acetone, and in high concentration of acidic solution at high temperature. Notably, the lignin is easily soluble in alkaline solutions at both low and high temperature [25]. In a general thermal temperature heating of range 150–300°C degrade hemicellulose, 315–400°C degrade cellulose and 250–500°C degrade lignin from plant biomass. In lignin structure, three aromatic phenolic components coniferyl (G), p-coumaryl (H), and sinapyl (S) are the major monomers (**Figure 3**).

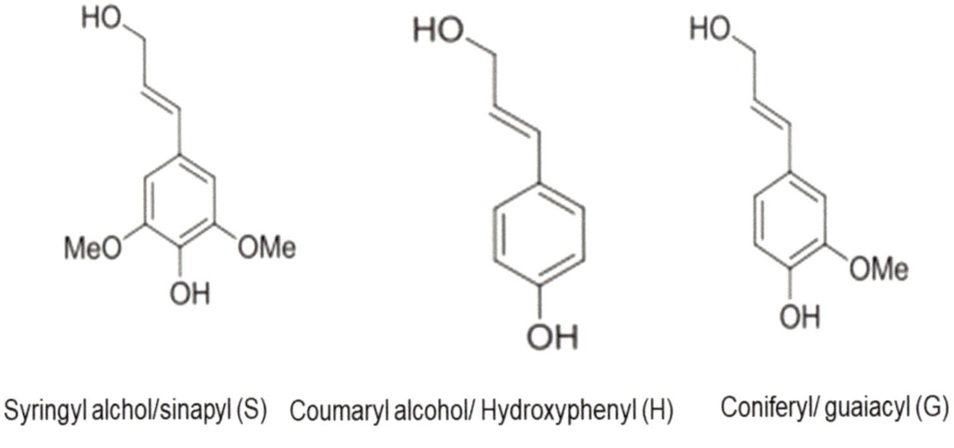

Figure 3.
Three main components in lignin structure.

Generally phenolic hydroxyl, methoxyl, and terminal aldehyde groups are present in the lignin. Softwoods lignin comprise of guaiacyl units, while in hardwoods the lignin contain guaiacyl-syringyl, whereas *p*-coumaryl alcohol units are found in Graminaceous lignin. In lignin synthesis, alcoholic hydroxyl and carbonyl groups are

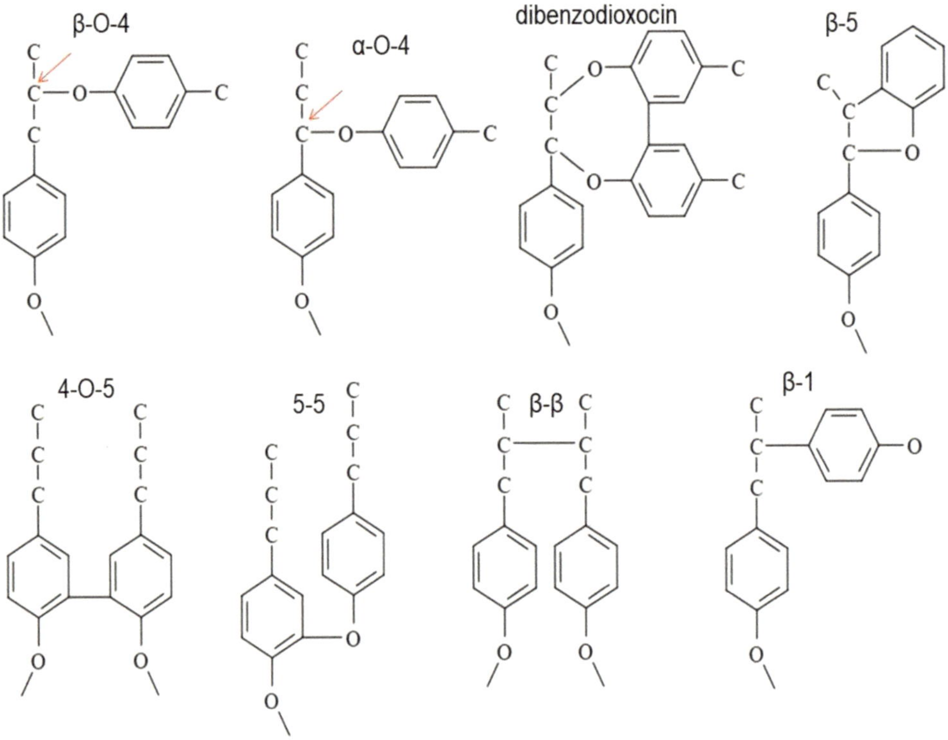

Figure 4.
Basic phenylpropane linkages in lignin structure.

assimilated into the structure of lignin during enzymatic dehydrogenation. Lignin structure has some mutual functional groups and common linkages between the aromatic phenolic components. Few of them are, β-O-4, α-O-4, dibenzodioxocin, β-5, 5–5, 4-O-5, β-β, and β-1 linkages. The density of the lignin structure is from the monomers ratio, compositions pattern and due to a number of linkages options between lignin monomers. Some of the common bonds arrange in carbon-carbon bonds (β-1, β- β, 5–5′) carbon-oxygen bonds (α-O-4, β-O-4, and 4-O-5), carbon-oxygen and carbon-carbon bonds (β-5′/ α-O-4, β- β′/α-O-γ). One the prominent 60% of total is ether β-O-4 linkage present in the polymer (**Figure 4**) [26].

3. Chemical groups in biomass

Plant biomass has various functional groups i.e. phenolic and furfural compounds, among them the important groups are listed in the (**Table 2**). Among cellulose, hemicellulose and lignin, highest different functional groups are present in lignin polymer. Hydrogen and ether glucosidic bonds are common in cellulose. Thus cellulose can be easily transformed and fractionated into sugars monomers. Kirk-Otmer [27] defined the alteration, substitutions and transformations in functional groups. In lignin cholrine and nitro groups are added in the substitution reactions of chlorination and nitration. Reaction like, acidic solutions change hydroxyl group to allylic or aryl ether in lignin and make it water soluble. The glucosidic linkage in sugars is the most important bond in ether groups, in lignin polymer. Removal of this ether bond distort the chain in lignin and start separation of lignin from the rest of plant biomass. Condition of alkaline solution reomve aromatic rings, whereas, solvolytic reaction split ether bond, and acidic reaction transform ether bond into hydroxyl and carboxyl prior fragmented into C3 or C2 molecules.

In hemicellulose, the common bond is hydrogen or ester bond with a hydroxyl chain. In long chain of hemicellulos polysaccharides, the ester bond connect to either carboxyl or hydroxyl groups but the position is not fixed between the cross linking of hemicellulose-lignin, and hemicellulose–cellulose [22]. The glucose monomer are connected by hydrogen bonds, oxygen atom and a hydroxyl group in hemicellulose and cellulose polymer chains. These bonding makes the polysaccharides in plant biomass rigid and only harsh conditions of high temperatures acids and alkali can break bonding of the polymer structure of the biomass [28].

Group	Cellulose	Hemicellulose	Lignin
Hydroxyl group	X	X	√
Carbon to carbon linkage	X	X	√
Ester bond	X	√	X
Hydrogen bond	√	√	X
Ether (glucosidic linkage)	√	√	√
Aromatic ring	X	X	√

Table 2.
Basic functional groups in plant biomass.

4. Lignin types, isolation and pretreatment methods

Base on the isolation and pulping process, lignin products is termed with five different names

1. Technical lignin = which is made of paper, pulp and cellulose at industrial level

2. Kraft lignin = containing sulfate after pulping process

3. Lignosulfonate = containing sulfite groups in pulping process

4. Alkali lignin = lignin isolated from biomass using different alkali

5. Organosolv lignin = isolated with organic solvents treatment

As it known to us that the lignocellulosic biomass has three major components i.e. lignin, hemicellulose and cellulose that are linked together through different linkages and bonds between them. The details components and linkages are shown in the **Figures 3** and **4**. The isolation of lignin requires to disrupt the covalent, ether and hydrogen bonds between lignin and other fraction. The efficacy of the process arises from the yield, purity, and the degree of transformation in lignin structure. In each isolation process, either full or partial cleavage of bonds and functional groups are targeted. During the isolation method, the lignin linkages can be exposed to changes such as ether, carbon-carbon, oxygen-carbon bonds alteration and other than degradation to small fractions. Different kinds of pretreatment methods (**Figure 5**) are tested to effectively separate out the main components of plant biomasses. The methods tested includes physicals, biologicals and chemicals to degrade and hydrolyze

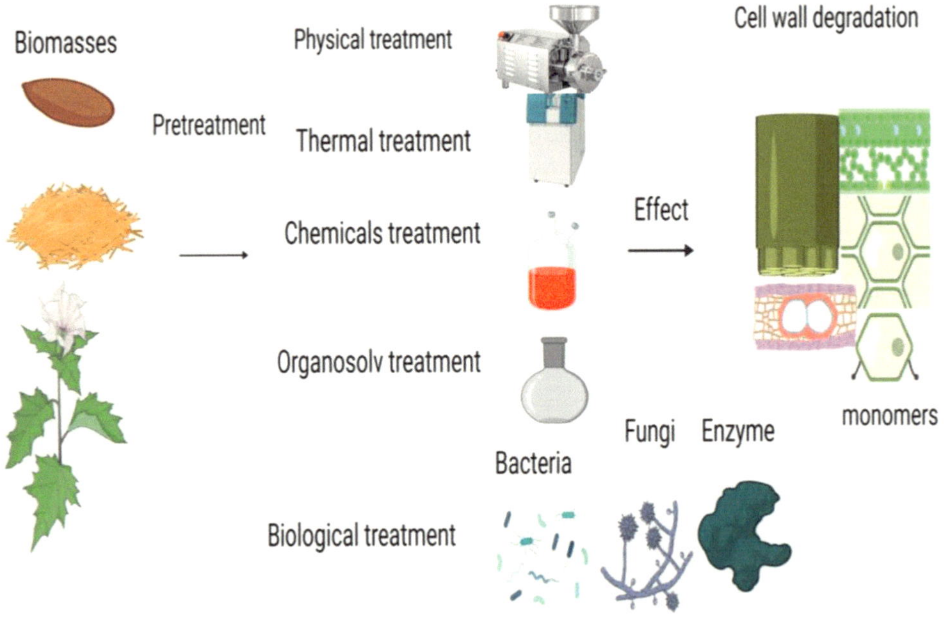

Figure 5.
Pretreatment methods and their effect on biomass.

the targeted biomasses into its monomers. The pretretments are necessory to remove lignin, one of the toughest structure that act as an obstacle for its utilization purposes. Thus lignin is considered to be remove from the biomasses and makes the cellulosic component expose for the transformation into usable materials [29].

The preeminent approach must have characteristics like (a) minimum chemicals uses (b) high carbohydrates reclamation (c) narrow quantities of by-products (d) economical far-reaching applications (e) appropriate for diverse biomasses and (f) decrease the concentration of enzymes used in hydrolysis of biomasses [30]. So far none of a single pretreatment method has all the good properties as supposed above, but in broad-spectrum, this section will concisely focus on the common methods reported for different biomass hydroylsis.

4.1 Mechanical treatment

Mechanical method change the solid fragments of the biomass into granular shape, improve the precise surface area and degrade into lesser particles, remove water content, decrease the needs high amount of energy consumption in further pre-treatment reaction. Various mechanical methods like high-pressure homogenizer, piston press, bead mill, grinding and sonication are reported.

4.2 Thermal method

A high temperature utilization is also applied at commercial scale that's disintegrate the plant biomass and direct conversion of biomasses into polysaccharides and carbon materials. The thermal treatment scale depends on the plant biomass materials and conversion into target molecules. Low thermal temperature ranges from 120 to 180°C are uses for proteins, carbohydrates and other sugars molecules solubilization. Wherase, high thermal treatment are applied for prouction of volatile organics, carbons and gases from biomass. Thermal heating in autoclave uses steam heat under high pressure on lignocellulosic biomass results in a considerable break in lignin linkages in biomass [31]. Electromagnetic radiation from microwaves with a frequency range of 250–300 MHz energy induced molecular vibration and is showed more efficient and less time consuming pretreatment method compared to conventional heating [32]. The microwave heating combine with acid, alkali and organic solvents has high impact than alone thermal treatment method [33]. So thermal treatment with any technology can be use to disintegrate the plant biomass into its components.

4.3 Lignosulfonate process

The isolation of lignin using sulfite and bisulfite anions at industrial scale. In this process, sulfur dioxide reacts with water with control oxygen supply in sulfur combustion and produce sulfurous acid for making pulping liquor. This reaction use alkali base or carbonates for pH (1–6) adjustment, and temperature adjustment of 140–160°C. The final product lignosulfonate polymer is water soluble and comprises of 5% sulfur and carbohydrate impurities.

4.4 Kraft process

The Kraft process is a universal delignification method used in paper and pulp industry for lignin removal. In this process sodium sulfide (NaS2) and sodium hydroxide NaOH are used to disassociate lignin linkages in substrates to separate out cellulose. The addition of these alkali solutions at high temperature of 150–170°C for about 100–120 minutes, results a brown liquid and solid pulp. The lignin is precipitated through acidification from dark liquor and purified while cellulose and other carbohydrate residues and inorganic impurities are left aside. Purified Kraft lignin is then used to energy purpose through combustion in paper mills.

4.5 Chemical method

Chemical method with thermal heating is commonly used for fractionation of biomass into lignin cellulose, and hemicellulose. The function of acids in low concentration break the outer surface and release bulk of extratives and hemicellulose of the biomass. Wheras, the concentrated acids at high thermal treatment disassociate the firm structure of the biomass. This treatment release both lignin and hemicellulose sugars of the biomass simultaniously. Sulfuric acid, hydrocholric acid, and nitric acid with (1–10% w/v) are mostly reported in the acid chemcial treatment method [34]. The alkalin solutions particularly of sodium hydroxide, potassium hydroxide, and calcium hydroxide specifical target linkages bonds in lignin and has very less penetration power for the cellulose of biomass. The alkaline solution with thermal heating pretreatment makes the biomass digestible for enzymes saccharification into sugars [35] allowing maximum entrance to hemicellulose and cellulose molecules [36]. It is hard to classify the most appropriate pre-treatment process for different kinds of biomasses [37]. But, the right choice pre-treatment process needs to enhance porosity of the biomass and reduce the inhibitors generation [37]. Ideally, low acids and alkal concentration has revealed optimmum separation of lignin, hemicellulose and cellulose of the biomass without addition of inhibitors [38]. Further, dilute acids and alkali pre-treatment can are economical for bioenergy and biochemicals production from biomass polysaccharides at large scale. Potassium hydroxide, calcium hydroxide and sodium hydroxide are the highly ested alkalies for lignocellulose degradation, however, sodium hydroxide has a greater potential reaction rate compare to other alaklies [39]. Wherase, in cost perspective, calcium hydroxide has less cost per kg than the other and could prove to the inexpensive pretreatment of biomass [40]. Therefore, for industrial scale process, it is better to select the inexpensive acids and alkali with optimum thermal process for yield of polysaccharides and biofuels production from pretreated biomass through anaerobic fermentation.

4.6 Organosolv method

Organosolv extraction method include, an organic solvent dissolved in water at high hydrothermal condition for biomass hydrolysis to remove a pure sulfur-free lignin. Common tested solvents have very low boiling point and can be simply recycle back after the pretreatment process. The subsequent lignin structure has small changes related to the native lignin. The organosolv treatments in fact chop favorably the bonds between carbohydrate-lignin molecules separating a partially modified lignin. Due to this characteristic, lignin obtained with organosolv treatments is a best monomer for making polymers, such as bio-based polymers, polyesters, polyurethanes, and phenolformaldehyde resins. The most universal organic solvents are acetone, ethanol, methanol, or a mixture thereof. The purified lignin is recuperated after the precipitation process using water.

4.7 Biotreatment

Biotreatment is basically biological pretreatment method applying either bacteria, fungi or yeast cells in sinngle culture of mix microbial communities composition. Additionally, specific enzymes in as a single entity or mixed enzymes solution has also been used in biotreatment system. The nature and function of enzymes and microbial culture could be different but the overall target is to fractionate the biomass into separate polysaccahrides molecules. A number of enzymes such as lignin peroxidases, laccases, xylanases, endoglucanases, peptidases, esterase and hydrolases has been applied in biomass saccharification process. Beside enzymes, a number of fungal strains *Phanerochaete chrysosporium, Trametes versicolor, Pleurotus ostreatus, Ceriporiopsis subvermispora,* and bacterial speies particularly *Clostridum sp. Pseudomonas sp.* LD002 [41], actinomycetes [42], *Streptomyces viridosporus* T7A [43], *Rhodococcus jostii* RHA 1 and *Pseudomonas putida* [44, 45], *B*acillus sp. strains [20–23], and thermophile *anaerocellum thermophilum* are reported for oxidative hydrolysis of lignin from biomass [46, 47]. Fungi like white and brown-rot fungi has the ability to hydrolyze both lignin and cellulose [48]. Such strains has the enzymes machinery of lignin peroxidase, manganese peroxidase, versatile peroxidase, laccase and cellulases [49]. The eficiency and percentage of lignin degradation depends on the enzymes and growth of the microorgansims like *Stereum hirsutum (*white-rot fungi) has 14.5%, degradation of the lignin from the wood biomass [49]. Wherase, *Coniochaet ligniaria* fungus reported upto 75% of lignin degradation from pepper plant, likwise, *Pleurotus Florida* with 45–50% degradation of lignin from corn straw [50]. The importance of biological treatment method is consumption of low energy, safe for environment and has no involvement of toxic chemcials. However, the problems that needs to be adjusted is sluggish growth, optimum conditions of enzymes, products inhibition, pretreatment time, enzymes purification, economical approach, feasible for the degradation of diverse agriculture biomasses [51]. However, all of these studies proposed that biological treatment is economical pre-treatment method among the all for inexpensive biofuels productions.

5. Lignin applications

Isolated lignin through any pretreatment methods can be use directly as unmodified crude lignin or pure lignin with further modification reactions in the applied field. This section will briefly describe the main applications of both native lignin and modified lignin.

5.1 Native lignin

Lignin has many applications; one of them is direct combustion of lignin as black liquor as a fuel and power in paper and wood pulping industry. Also the native lignin increase combustion heat to bio-fire made of propane-1,3-diol and cellulose. In view of its binding stuffs, it is commonly used as a stabilizer in coal briquettes in which it raises the boiling speed and strength in packing paper material. Regardless of the importance of lignin for power generation, it has many other uses like as toluene or benzene, non-toxic binder in pelleted feeds, nonhazardous chelating agent, transporter of nutrients as lignosulfonate for the plants to discharge them gradually in the soil. Lignin can also be use as a sequestrant of heavy metals (e.g. zinc, nickel, cadmium and mercury) for purification of polluted water due to chelating property [26, 52].

5.1.1 Polymers from native lignin

Native lignin can be an outstanding preparatory material for making of polymer composites with other molecules through covalent bonds. Additionally, as an additive the native lignin can protect polymers from high temperature, light and oxidation. The most essential linkages are the hydroxyl groups in polymerization. Remarkably, the steric interruption decrease the reactivity of lignin and due to fairly small amount of free active hydroxyls, still the use of lignin in polymer materials has been successful. For example, in the reaction between phenol and formaldehyde for production of phenol-formaldehyde adhesives of chipboards, plywood etc., lignin is convenient as a substitute for phenol binding reagent. Similarly, lignin can be blended in polyethylene, polypropylene, to change its mechanical properties and improving the oxidative resistance. Kraft-lignin as an unmodified material used in polyesters as a reaction monomer has been reacted with sebacoyl chloride, increased the recycling of subsequent polymer. Native lignin has many other applications also, it can react with polyethylene glycol (PEG) and methylene diphenyl isocyanate to make polyurethanes, biodegradable foam, cheap composites, bioplastic materials mixed with cellulose fibers, cellulose-lignin based hydrogel, medical devices or controlled drug release exercises [26, 52].

5.2 Modified lignin

Lignin reaction can be change and modified to add new functionalities with aim for fine chemicals production.

5.2.1 Adding new function

Amination, a reaction to add reactive amino group to lignin to synthesize aminated lignin hat has significant application as a slow-release mediator for fertilizers and flocculant to remove anionic dyes from water solutions [26].

5.2.2 Synthesis of polymers

The native lignin can be used for synthesis of several polymers. To do this, changes in the reactive sites such as phenol and hydroxyl groups like lignin-phenolation is the mostly used process. Lignin-phenolation is carried out using methanol or ethanol in acidic medium, with a boiling point temperature of the tested solvent to make phenol-formaldehyde (PF) resins. Similarly, sulfur-mediated demethylation is a process to remove the methyl from the hydroxyl groups in lignin is economical and most significant reaction to synthesize phenol-formaldehyde resins [26].

5.2.3 Catalytic reactions

There are feasibly many catalytic reactions for lignin modification; but the three important of them are oxidation, hydrogenolysis, and hydrodeoxygenation.

5.2.4 Oxidation

In oxidation, ether bonds are sliced of the hydroxyl-benzyl group of the lignin structure, resulting to a combination of aromatic ketones and aldehydes that are valuable predecessors to fine chemicals. For example, in an alkaline medium, lignin is oxidized under high-pressure of oxygen for vanillin synthesis. Further, catalysts like Pd/Al2O3 speed up the selective oxygenation of lignin for high yield of vanillin and syringaldehyde [26].

5.2.5 Hydrogenolysis

Hydrogenolysis is a reaction that breaks ether bonds in lignin at high temperature and hydrogen in the presence of a catalyst. For example, in the presence of Ni, lignin is converted into aromatics compounds such as dihydroconiferyl alcohol, dihydroxysinapyl alcohol, guaiacol, and 4-propylsyringol. In hydrogenolysis, the selective reaction on specific bonds of the lignin towards valuable chemicals, identification and separation of the specific product from the complex mixture of aromatics is a challenging task [26].

5.2.6 Hydrodeoxygenation

Hydrodeoxygenation is a degradation reaction that gives rise to aromatics with low oxygen content. In hydrodeoxygenation, under strict temperature and pressure using a catalyst, the lignin can be converted into benzene, cycloalkanes and hydrogenated bio-oil as a fuel [26]. A list of lignin base products synthesized through different reaction process are given below in **Table 3**.

The high abundance and availability of lignin has a great market potential for industrial use. In future the efficient utilization of the isolation process and chemical conversion reaction with new technologies may have significant contributions

Product	Lignin or Lignin Monomers Reaction
Feruloyl and p-coumaryl	Conversion of ferulate and coumarate by Acyl-CoA synthetase (HcaC) of Acinetobacter baylyi, Cupriavidus necator JMP134
Vanillin	Conversion of feruloyl-by Acyl-CoA hydratase (HcaA) of A. baylyi
Phydroxybenzaldehyde	onversion of p-hydroxybenzaldehyde by Dehydrogenase (HcaB) of A. baylyi,
Guaiacol	Vanillate by Decarboxylases (VdcB, C, D) of Streptomyces sp. D7
Formaldehyde	Methyl-THF by O-Demethylase systems (MetF) of Acetobacterium dehalogenans, Sphingobium sp. SYK-6
Methanol and C8–C9 alkanes	Conversion of Birch wood sawdust at 200/250°C, 40 bar, Ru/C, Pd/C, Rh/C, Pt/C
Sugars	Wood flours from Japanese beech (Fagus crenata) using catalyst 1 ethyl-3-methylimidazolium chloride at 90–120°C
Phenol	Sugarcane bagasse lignin using Raney Ni and β-zeolite at 2 h at160 oC; 2 h at 240°C conditions
Plywood and fiber board binder	Making of binder of through phenolation that condense lignin aromatics monomers with phenol

Product	Lignin or Lignin Monomers Reaction
Adhesives	Reaction of lignin with phenolic resin to make phenol-formaldehyde (PF) adhesive
Asphalt	Bitumen in asphalt can be substituted by lignin to make epoxy concrete or epoxy-cement.
Bio-sourced battery	In lithium-ion batteries for enhanced electrical systems lignin can bind cathode, electrolyte, and anode as a bio-sourced polymer
Lignosulfonates	lignosulfonates of sulfite spent liquors (SSL), calcium, and sodium can produce wood adhesives
Medical/healthcare devices	plant-based adhesive hydrogel using Ag-Lignin nanoparticles (NPs), The hydrogels is made of a solution of poly (ethylene glycol), acrylic acid (AA), pectin, ammonium persulfate, Ag-Lignin NPs solution, and diacrylate.
Biocomposites	treatment of cotton fiber or polylactic acid (PLA) as a matrix for the composites with native lignin extracted from biomass
Leather skeleton	Lignin coated leather can replace using azobenzene in dyeing industries
Lignin coating polymer	Klason lignin treatment with hexamethoxy methyl melamine, p-toluene sulfonic acid goes for self-polyesterification with sodium hypophosphite (NaHP) to produced polymer

Table 3.
Examples of few important lignin base products [26, 52].

for chemicals, materials and fuels from raw lignocellulosic biomass. The successful application of novel process will build a sustainable route in future bioeconomy. In brief the lignin can be used to be converted through the reactions described in to many chemicals products. The lignin has valuable raw materials for several applications. The potential products from lignin for the market are

- Syngas products (Dimethyl ether, Methanol, Ethanol, C1–C7 gases)
- Hydrocarbons (Benzene, toluene, xylene, higher alkylates, cyclohexane, styrenes, biphenyls)
- Phenols (Phenol, catechols, cresols, resorcinols, syringols, coniferols, guaiacol)s
- Oxidized products (Vanillin, vanillic acid, syringaldehyde, syringic acid, aliphatic acids, cyclohexanol/al, β-keto adipate)
- Macromolecules (Carbon fibers, polymer alloys, composites, adhesives, binders, preservatives, polyols)

6. Conclusion

As global fossil resources are immensely diminishing and high utilization of it causing climate pollution due to discharging of harmful gases. Lignin was assumed as useless materials in plant biomasses and same for other waste residue of lignocellulosic substrates. So a huge amount of beneficial carbon materials vanished either burning in the field or for cooking purpose. This chapter highlights that plant biomass contains cellulose, hemicellulose and lignin. All of the three main components can be convert into industrially valuable products. This chapter described the accessibility of huge amount of lignin along with cellulose and hemicellulose carbons, separation methods of polysaccharides and conversion of these usable carbons to high value chemicals production. This brief information about lignin has the possibility to be extended for better depolymerization of lignin base biomasses, and a knowledge source for efficient reaction to new aromatic catabolic chemicals in the future.

Acknowledgements

The author of this book chapter acknowledge Prof. Li Zhihe and Dr. Li Zhiyu for reviewing the contents and feedbacks. The author is also thankful to School of Agriculture Engineering and Food Sciences at Shandong University of Technology China for providing working facilities.

Conflict of interest

The authors declare that they have no known competing financial interests or personal relationships that could have appeared to influence the work.

Notes/thanks/other declarations

The author has no other declarations, and data sharing is not applicable to this article as no new data were created in this study.

Author details

Tawaf Ali Shah*, Li Zhihe, Li Zhiyu and Zhang Andong
School of Agriculture Engineering and Food Science, Shandong University of Technology, China

*Address all correspondence to: tawafbiotech@yahoo.com

References

[1] Ali S, Shah TA, Afzal A, Tabassum R. Exploring lignocellulosic biomass for bio-methane potential by anaerobic digestion and its economic feasibility. Energy & Environment. 2018;**29**(5):1-10

[2] Gellerstedt G, Henriksson G. Lignins: major sources, structure and properties. Monomers, Polymers and Composites from Renewable Resources. 2008;**9**(4):201-224

[3] Mutturi S, Palmqvist B, Lidén G. Developments in Bioethanol Fuel-Focused Biorefineries. Advanc Biorefiner: Elsevier; 2014. pp. 259-302

[4] Khan A, De Jong W, Jansens P, Spliethoff H. Biomass combustion in fluidized bed boilers: Potential problems and remedies. Fuel Processing Technology. 2009;**90**(1):21-50

[5] Shah TA, Asifa A, Shehbaz A, Rahim U, Romana T. A review on biohydrogen as a prospective renewable energy. International Journal of Bioscience. 2017;**11**(1):106-130

[6] Krausmann F, Erb K-H, Gingrich S, Lauk C, Haberl H. Global patterns of socioeconomic biomass flows in the year 2000: A comprehensive assessment of supply, consumption and constraints. Ecological Economics. 2008;**65**(3):471-487

[7] Shah TA, Ali S, Afzal A, Tabassum R. Simultaneous pretreatment and biohydrogen production from wheat straw by newly isolated ligninolytic *bacillus sp. strains* with two-stage batch fermentation system. BioEnergy Research. 2018;**11**(4):835-849

[8] Aimen S, Hafiz U, Shah TA, Tabassum R. Fermentation of simple and complex substrates to biohydrogen using pure *Bacillus cereus strains*. Environment Technology & Innovative. 2020;**18**(1):100704

[9] Shah A, Favaro L, Alibardi L, Cagnin L, Sandon A, Cossu R, et al. *Bacillus sp. strains* to produce bio-hydrogen from the organic fraction of municipal solid waste. Applied Energy. 2016;**176**:116-124

[10] Shah TA, Raheem U, Asifa A, Romana T. Effect of alkalis pretreatment on lignocellulosic waste biomass for biogas production. International Journal of Renewable Energy Research (IJRER). 2018;**8**(3):1318-1326

[11] Hoekstra AY, Wiedmann TO. Humanity's unsustainable environmental footprint. Science. 2014;**344**(6188): 1114-1117

[12] Nakićenović N, Grübler A, McDonald A. Global energy perspectives. Vol. 6. No. 98. Cambridge: Cambridge University Press; 1998. pp. 462-470

[13] Pérez-Lombard L, Ortiz J, Pout C. A review on buildings energy consumption information. Energy and Buildings. 2008;**40**(3):394-398

[14] Shah TA, Ullah R, Muhammad M, Rashida. Reprocessing of NaOH black liquor for pre-treatment of agribiomass. International Research Journal of Agriculture Science Veterinary Medicine (IJAVM). 2020;**8**(4):1-10

[15] Shah TA, Tabassum R. Enhancing biogas production from lime soaked corn cob residue. International Journal of Renewable Energy Research (IJRER). 2018;**8**(2):761-766

[16] Kumar R, Singh S, Singh OV. Bioconversion of lignocellulosic biomass:

Biochemical and molecular perspectives. Journal of industrial Microbiology & Biotechnology. 2008;**35**(5):377-391

[17] Ali S, Shah TA, Afzal A, Tabbassum R. Evaluating the co-digestion effects on chicken manure and rotten potatoes in batch experiments. International Journal of Bioscience. 2017;**10**:150-159

[18] Shah TA, Ullah R. Pretreatment of wheat straw with ligninolytic fungi for increased biogas productivity. International Journal of Environment Science Technology. 2019:**16**:7497-7508

[19] Afzal A, Shah TA, Saleem M, Tabassum R. Characterization of extracellular protease from *bacillus licheniformis*. International Journal of Biosciences. 2018;**11**(4):228-236

[20] Shah TA, Lee C, Orts WJ, Tabassum R. Biological pretreatment of rice straw by ligninolytic *bacillus sp. strains* for enhancing biogas production. Renewable and Sustainable Energy Reviews. 2018;**38**(3):1-9

[21] Sommer SG, Hamelin L, Olesen JE, Montes F, Jia W, Chen Q, et al. Agricultural waste biomass. Supply Chain Management for sustainable Food Networks. 2016:67-106

[22] Faulon J-L, Carlson GA, Hatcher PG. A three-dimensional model for lignocellulose from gymnospermous wood. Organic Geochemistry. 1994;**21**(12):1169-1179

[23] Anwar Z, Gulfraz M, Irshad M. Agro-industrial lignocellulosic biomass a key to unlock the future bio-energy: A brief review. Journal of Radiation Research and Applied Sciences. 2014;**7**(2):163-173

[24] Menon V, Rao M. Trends in bioconversion of lignocellulose: Biofuels, platform chemicals & biorefinery concept. Progress in Energy and Combustion Science. 2012;**38**(4): 522-550

[25] Harmsen P, Huijgen W, Bermudez L, Bakker R. Literature review of physical and chemical pretreatment processes for lignocellulosic biomass. Wagen UR-Food & Biobas Research. 2010;**54**(1):1184

[26] Abdelaziz OY, Brink DP, Prothmann J, Ravi K, Sun M, García-Hidalgo J, et al. Biological valorization of low molecular weight lignin. Biotechnology Advances. 2016;**34**(8):1318-1346

[27] Kirk TK, Farrell RL. Enzymatic combustion: The microbial degradation of lignin. Annual Reviews Microbiology. 1987;**41**(1):465-501

[28] Bochek A. Effect of hydrogen bonding on cellulose solubility in aqueous and nonaqueous solvents. Russian Journal of Applied Chemistry. 2003;**76**(11):1711-1719

[29] Cragg SM, Beckham GT, Bruce NC, Bugg TD, Distel DL, Dupree P, et al. Lignocellulose degradation mechanisms across the tree of life. Current Opinion in Chemical Biology. 2015;**29**:108-119

[30] Biswas R, Uellendahl H, Ahring BK. Wet explosion: A universal and efficient pretreatment process for lignocellulosic biorefineries. Bioenergy Research. 2015;**8**(3):1101-1116

[31] Ruiz E, Cara C, Manzanares P, Ballesteros M, Castro E. Evaluation of steam explosion pre-treatment for enzymatic hydrolysis of sunflower stalks. Enzyme and Microbial Technology. 2008;**42**(2):160-166

[32] Zhu S, Wu Y, Yu Z, Liao J, Zhang Y. Pretreatment by microwave/alkali of rice straw and its enzymic hydrolysis. Process Biochemistry. 2005;**40**(9):3082-3086

[33] Zhu S, Wu Y, Yu Z, Chen Q, Wu G, Yu F, et al. Microwave-assisted alkali pre-treatment of wheat straw and its enzymatic hydrolysis. Biosystems Engineering. 2006;**94**(3):437-442

[34] Lenihan P, Orozco A, O'neill E, Ahmad M, Rooney D, Walker G. Dilute acid hydrolysis of lignocellulosic biomass. Chemistry Engineer Journal. 2010;**156**(2):395-403

[35] Yan Z, Li J, Chang S, Cui T, Jiang Y, Yu M, et al. Lignin relocation contributed to the alkaline pretreatment efficiency of sweet sorghum bagasse. Fuel. 2015;**158**:152-158

[36] Song Z, Liu X, Yan Z, Yuan Y, Liao Y. Comparison of seven chemical pretreatments of corn straw for improving methane yield by anaerobic digestion. PLoS One. 2014;**9**(4):e93801

[37] Hahn-Hägerdal B, Galbe M, Gorwa-Grauslund MF, Lidén G, Zacchi G. Bio-ethanol–the fuel of tomorrow from the residues of today. Trends in Biotechnology. 2006;**24**(12):549-556

[38] Hu F, Ragauskas A. Pretreatment and lignocellulosic chemistry. Bioenergy Research. 2012;**5**(4):1043-1066

[39] Gupta R, Lee Y. Pretreatment of corn Stover and hybrid poplar by sodium hydroxide and hydrogen peroxide. Biotechnology Progress. 2010;**26**(4):1180-1186

[40] Kootstra AMJ, Beeftink HH, Scott EL, Sanders JP. Comparison of dilute mineral and organic acid pretreatment for enzymatic hydrolysis of wheat straw. Biochem Engineer Journal. 2009;**46**(2):126-131

[41] Fong S, S, B Fisher A. Lignin biodegradation and industrial implications. AIMS Bioengineer. 2014;**1**(2):92-112

[42] Tuomela M, Vikman M, Hatakka A, Itävaara M. Biodegradation of lignin in a compost environment: A review. Bioresource Technology. 2000;**72**(2):169-183

[43] Crawford DL, Pometto AL, Crawford RL. Lignin degradation by *Streptomyces viridosporus*: Isolation and characterization of a new polymeric lignin degradation intermediate. Applied and Environmental Microbiology. 1983;**45**(3):898-904

[44] Ahmad M, Roberts JN, Hardiman EM, Singh R, Eltis LD, Bugg TD. Identification of DypB from *Rhodococcus jostii* RHA1 as a lignin peroxidase. The Biochemist. 2011;**50**(23):5096-5107

[45] Bugg TD, Ahmad M, Hardiman EM, Singh R. The emerging role for bacteria in lignin degradation and bio-product formation. Current Opinion in Biotechnology. 2011;**22**(3):394-400

[46] Chang J-J, Chou C-H, Ho C-Y, Chen W-E, Lay J-J, Huang C-C. Syntrophic co-culture of aerobic *bacillus* and anaerobic *clostridium* for bio-fuels and bio-hydrogen production. International Journal of Hydrogen Energy. 2008;**33**(19):5137-5146

[47] Chang T, Yao S. Thermophilic, lignocellulolytic bacteria for ethanol production: Current state and perspectives. Applied Microbiology and Biotechnology. 2011;**92**(1):13-27

[48] Galbe M, Zacchi G. Pretreatment of Lignocellulosic Materials for Efficient Bioethanol Production. Biofuels: Springer; 2007. pp. 41-65

[49] Lee J, Gwak K, Park J, Park M, Choi D, Kwon M, et al. Biological

pretreatment of softwood Pinus densiflora by three white rot fungi. Journal of Microbiology Seoul. 2007;**45**(6):485

[50] Zhong W, Zhang Z, Qiao W, Fu P, Liu M. Comparison of chemical and biological pretreatment of corn straw for biogas production by anaerobic digestion. Renewable Energy. 2013;**51**:518

[51] Lynd LR, Van Zyl WH, McBride JE, Laser M. Consolidated bioprocessing of cellulosic biomass: An update. Current Opinion in Biotechnology. 2005;**16**(5):577-583

[52] Mili M, Hashmi SAR, Ather M, Hada V, Markandeya N, Kamble S, et al. Novel lignin as natural-biodegradable binder for various sectors—A review. Journal of Applied Polymer Science. 2022;**139**(15):51951

Chapter 3

Perspective Chapter: Potential of Lignin Valorization with Emphasis on Bioepoxy Production

Mohamed H. El-Moayed, Justin Kühn, Sea-Hyun Lee, Mahmoud Farag and Sherif Mehanny

Abstract

Lignin is the second most abundant natural polymer after cellulose. It has high molecular weight and poor dispersity, which lowers its compatibility with other polymeric materials. Accordingly, it is hard to integrate lignin into polymer-based applications in its native form. Recently, lignin valorization, which aims to boost lignin value and reactivity with other materials, has captured the interest of many researchers. The volatility of oil and gas prices is one strong incentive for them to consider lignin as a potential replacement for many petroleum-based materials. In this chapter, lignin valorization processes, namely hydrogenolysis, pyrolysis, hydro-thermal liquefaction, and hydro-thermal carbonization, are discussed in brief. The chapter also discusses the synthesis of lignin-based epoxy resin as an already existing example of a lignin-based product.

Keywords: lignocellulose, valorization, biomass, waste management, bioresins, eco-friendly

1. Introduction

Environmental friendliness and green alternatives have been a significant concern, to reduce carbon footprint and alleviate environmental toxicity problems [1, 2]. Lignocellulose is the most abundant renewable biopolymer ever existing [1, 3]. Cellulose, hemicellulose, and lignin count up to more than 90–95% of lignocellulosic texture [4]. Lignin is an organic polymer whose structure is highly dependent on its native source. In general, it is an amorphous, isotropic material, covered, for example, in the wooden stem of a plant. All plants and algae comprise lignin in different amounts, in addition to cellulose and hemicellulose [5, 6]. Lignin amount also differs according to the time of harvesting. The lignin content gets much higher, the older the plant gets. For instance, lignin presents about 12.4 ~ 29.4% of hemp [7] and 2.5 ~ 3.8% of flax [8]. Nevertheless, the highest lignin content can be found in woods, with a percentage of 20 ~ 30% [9]. Lignin leads to a so-called lignification of the stem (from Latin: *lignum*). This makes it more robust and stronger by gumming the cellulose part together [10]. This partly also occurs in leaves. It is extracted from these natural sources via physical, chemical, or biological methods.

Physical processes include steam explosion and mechanical grinding. The physical route yields high-purity lignin. However, it is hard to be industrially upscaled. Chemical processes commonly used for pulp and paper, such as the kraft and sulfite process, utilize reagents that trigger reactions, yielding moderate-purity lignin, depending on the parameters. The harsher the reaction environment and chemicals utilized, the lower the quality of lignin will become. New processes such as organosolv process running on water and ethanol as main solvents with an optional acid catalyst will result in higher quality lignin due to the mild reaction conditions. However, to obtain high-purity lignin, the respective fraction will need to undergo further treatment. Finally, the biological option involves enzymes that break lignin bonds with cellulose and hemicellulose. Despite producing high-purity lignin, this technique is not favorable due to its low speed. The majority of industrial lignins are extracted chemically from their sources.

Extracted lignin has high molecular weight and poor reactivity with other polymeric materials. Hence, its uses are limited to combustion applications. Around 98% of lignin is burnt as a low-value fuel, while the rest is fabricated into commercial products [11]. In order to increase lignin utility, its structure has to be modified first. The purpose of the modification is to enhance lignin reactivity and homogeneity and lower the probability of infusible solids formation. Boosting lignin value is called "lignin valorization." **Table 1** summarizes different valorization processes, which are covered in the following section.

With a share of 4.3% of the European pulp production and total production of 5 million tonnes of paper in Austria, there is an approximate market demand of 116.3 billion tonnes of pulp only in Europe. Multiplying this with a mean value of the content of lignin in wood, taking 25%, there is a total yearly amount of 29 million tonnes of lignin. Adding further lignin sources as mentioned above as well, the value is significantly increasing. The following chapter addresses, how the velarization process takes place to not waste this amount of lignin. Afterward, Chapter 3 deals

Process	Temperature range, °C	Atmosphere and media	Catalyst	Main products	Reference
Hydrogenolysis	40–200	H_2 water (or alcohol)	Supported transition metal	Mixed aromatics	[12]
Pyrolysis	500–1000	N_2, CO_2	Acid, alkali, metal, metal oxide	Pyrochar and bio-oil	[13]
Hydrothermal liquefaction	200–400	N_2, CO_2 Water (or Organic Solvent)	Acid, alkali, metal, metal oxide	Bio-oil and phenolics	[14]
Hydrothermal carbonization	160–240	N_2 water (or organic solvent)	Acid, alkali	Hydrochar	[15]

Table 1.
Lignin valorization processes.

with further processing of lignin-based epoxy resin synthesis. Concluding, a short chapter will summarize the findings.

2. Lignin valorization processes

To describe how to use lignin as a valuable resource (see **Figure 1**), this chapter initially describes the hydrogenolysis of lignin. Afterward, the pyrolysis and gasification are described that lignin might be used as an energy source. Material collection on a wet basis is discussed in the chapter on hydrothermal liquefaction (HTL). This chapter closes with a discussion of hydrothermal carbonization (HTC) of lignin.

2.1 Lignin hydrogenolysis

Lignin hydrogenolysis is a three-step process. First, lignin macromolecule functional groups are reduced without breaking the main structure. Second, macromolecules are broken into phenolics and arenes. Third, further reduction of second-step molecules into alkanes takes place.

2.2 Lignin pyrolysis and gasification

Lignin pyrolysis is an oxygen-free thermochemical conversion process. It aims to recover energy and other materials under a temperature ranging from 300 to 800°C [16]. Typical products of pyrolysis are biochar, bio-oil, and syngas. The quality of these products is highly dependent on temperature and dwell time. Longer residence time, or slower heating rate, reduces the bio-oil yield and boosts syngas content due to secondary cracking [17]. Lignin molecular weight is another important factor. Low molecular weight leads to the formation of CH_4, CO, and CO_2 from the methoxy group, whereas high molecular weight yields guaiacol and alkyl guaiacol [18].

While lignin pyrolysis is working in an oxygen-free environment for the production of syn-gas as the main product, which could also be of interest for future industries, as over 50% of syn-gas production is still petro-based and is based on the injection of pure oxygen or steam for stoichiometric balance, crucial for the

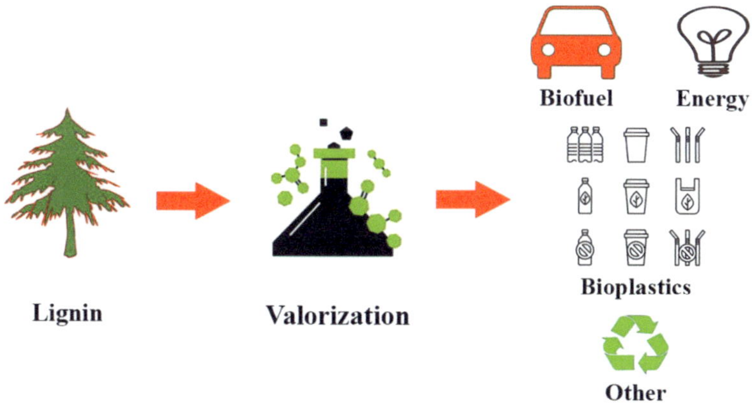

Figure 1.
Lignin valorization process.

production of higher quality syn-gas from biomass. The constitution of syn-gas is highly dependent on the biomass utilized, gasifier type, and agent, as well as operational conditions. It should be noted that lignin in itself is limited in hydrogen that is covalently bound to the structure ranging in the region of 5–6 wt%, which would be the maximum theoretically achievable ceiling, with respect to the energetic input needed for extraction. With gasification utilizing stream, the hydrogen output at higher temperatures is influenced due to the reaction between water and free carbon in the reaction chamber. Currently, the gasification of wood as the main feedstock is fairly common. However, the true interesting contender is biomass based on agricultural residues, such as straw, husks, shells, and more, which exhibit comparable values in syngas composition from the current benchmark in lab scale [19, 20].

2.3 Lignin hydrothermal liquefaction, HTL

Recently, HTL has been extensively studied as a biomass conversion method with water as a reaction medium. HTL is an efficient process of lignin transformation during which pyrolysis and hydrolysis take place simultaneously [14]. HTL is considered an eco-friendly and sustainable technology with the advantages of short residence time, high conversion rate, and less pollution.

Although lignin is abundant in plethoric quantities naturally, HTL has not reached the industrial scale, not even the pilot level because of troublesome depolymerization and separation. In contrast, lab-scale studies have been successful in lignin transformation in which 60 wt% low-weight aromatic molecules were obtained from oxidized lignin [21].

2.4 Lignin hydrothermal carbonization, HTC

HTC is an eco-friendly, convenient, and low-cost method [22]. Comparably, HTC offers distinctive advantages, which are not limited to no need for pre-drying, low processing temperature (180–350°C), avoiding air-polluting nitrogen oxides, and sulfur oxides dissolution in water [23]. During HTC operation, lignin is heated in a high-pressure autoclave in the presence of subcritical water and pressure ranging between 2 and 10 MPa. Reaction mechanisms comprise hydrolysis, dehydration, recondensation, decarboxylation, and aromatization [15], which are collaborative and concurrent. Similar to pyrolysis, key roles are played by the reaction temperature, residence time, and source of raw materials in the HTC process that yields hydrochar [24].

3. Lignin-based epoxy resin synthesis

Epoxy resins were first introduced in Europe by P. Schlack in 1939 [25]. They are a group of thermosetting resins (see **Figure 1**) that require curing to harden. The ratio of the curing agent, or the hardener, to the resin affects the overall performance of the polymer [26]. In general, epoxy possesses excellent thermal and mechanical properties. Hence, it is favorable in the fields of coating, electronic packaging, and thermal insulation [27], to name a few.

One of the most common epoxy resins is the diglycidyl ether of bisphenol A (DGEBA). DGEBA is the product of an epichlorohydrin (ECH) reaction with bisphenol A (BPA) in the presence of a basic catalyst [26]. Since BPA is a petroleum-based chemical, DGEBA is not biodegradable, thereby having a great potential to damage

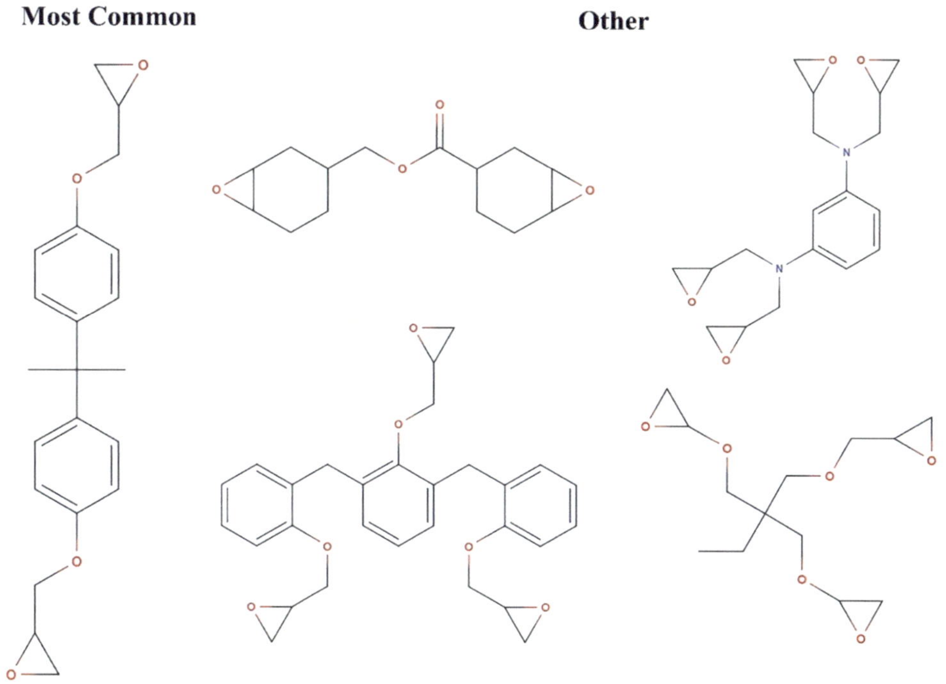

Figure 2.
Different types of epoxy resins.

the environment [28]. Being an environmental hazard has never been enough of an incentive to search for safer options. However, the recent volatility of oil and gas prices adds to the list of reasons that justify those options [29].

Lignin is a natural polymer whose structure is a complex network of phenyl-propane units, namely *p*-coumaryl alcohol, coniferyl alcohol, and sinapyl alcohol [29, 30]. **Figure 2** shows the three building blocks of lignin. Lignin structure is very similar to that of BPA, which qualifies the former to be an excellent substitute for the latter [28]. Luckily, lignin is the second most abundant macromolecule after cellulose. Therefore, it is no surprise that researchers find lignin-based epoxy resins worthy of their attention. By far, there are three methods to integrate lignin into epoxy resin synthesis. They are a) physical blending of lignin and epoxy resin, b) epoxidation of lignin after pretreatment, and c) epoxidation of unmodified lignin [31, 32].

3.1 Physical blending of lignin and epoxy resin

In this method, lignin is blended with petroleum-based epoxy resin to form a binary mixture, which is then cured at a proper temperature. *Simionescu et al.* [33] stated that epoxy resins with 25 ~ 50 wt% lignin exhibited good mechanical and dielectric properties after curing at elevated temperatures. In a different study, *Behin et al.* [34] reported that adding small amounts (< 2.5 wt%) of kraft lignin and Sal-A nanoparticles to uncured epoxy resins boosted their overall mechanical performance after cross-linking. They also mentioned that the additives had a positive impact on the curing reaction. In particular, both peak temperature and total heat of the curing reaction dropped significantly.

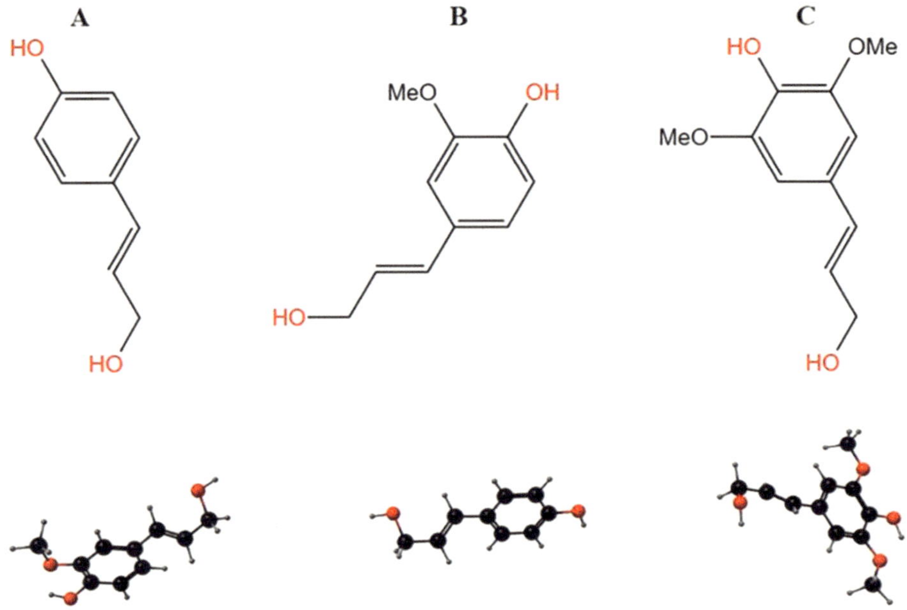

Figure 3.
Lignin's basic components: p-coumaryl alcohol (A), coniferyl alcohol (B), and sinapyl alcohol (C).

Pan et al. [35] considered the addition of lignin as a cross-linker. They mixed aminated lignin (see **Figure 3**) with a regular curing agent (W93) at 80°C. Then, a proper amount of liquid epoxy was added to the blend after cooling to room temperature. The epoxy-hardener mixture was cured in a baking oven afterward. Results showed that the thermal behavior of epoxy gradually improved with increasing the amount of aminated lignin in the hardener system up to 50%. Further increase in lignin content caused a drastic deterioration in performance, owing to agglomeration.

In light of the abovementioned studies, it is clear that the simple blending approach is straightforward and versatile. However, owing to lignin heterogeneity (steric-hindrance effect), the complete substitution of petroleum-based materials is not feasible with this technique.

3.2 Epoxidation of lignin after pretreatment

Large molecular weight and poor dispersity lower lignin compatibility with polymeric compounds. Better reactivity can be realized via chemical modification, for example, phenolation, hydroxymethylation, and demethylation. For instance, *Zhang et al.* [36] synthesized phenolated lignin, hydroxymethylated lignin, and demethylated lignin from native lignin through chemical modification. The modified lignin types were reacted with epichlorohydrin to obtain lignin-based epoxy resins, which were then utilized as feedstocks for new grouting materials. The newly prepared lignin-epoxy grouts outperformed the commercial ones, displaying great potential in replacing petroleum-based materials.

Zhen et al. [28] proposed a novel strategy (see **Figure 4**) to produce lignin-based epoxy resin from phenolated lignin. They demonstrated in their work the superiority of their new approach over the traditional one. They noticed a slight deterioration

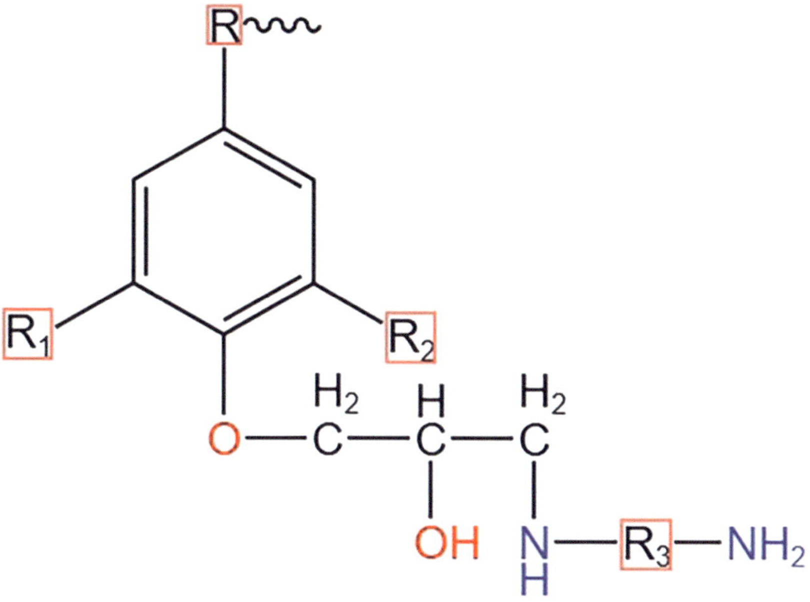

Figure 4.
Aminated lignin structure.

in thermal-mechanical performance when the lignin amount exceeded 20% (80% phenol). However, the values were still better than lignin-free resins (**Figure 5**).

Similar to the first approach, the steric-hindrance effect of lignin is still inevitable. However, its impact is not as bad as in the simple blending approach. With the technique discussed herein, the complete substitution of petroleum-based materials is achievable, which is a major step in the right direction toward the production of petroleum-free thermosets.

3.3 Epoxidation of unmodified lignin

Unmodified, or technical, lignins refer to a large group of lignin by-products whose properties are very different from their native form [37]. Kraft lignin, lignosulfonate lignin, and organosolv lignin are some good examples of unmodified lignins [38].

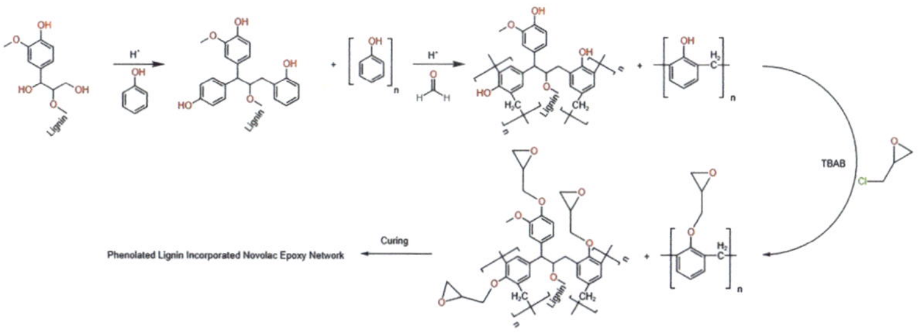

Figure 5.
PLIEN production process as proposed by Zhen et al. [28].

Like the previous method, BPA can be completely replaced with unmodified lignins in epoxy resin formulation. The main difference between the two techniques is the cost. Since technical lignins are already available as industrial wastes, no additional processing cost is added, which qualifies the current method for large-scale production [32].

4. Conclusions

The current chapter reviewed the current state of the art in lignin valorization starting from the extraction process to utilization via lignin-based epoxy resin synthesis. The key conclusions are as follows:

- While about every plant-based source has lignin as a base building material, the constitution will vary greatly depending on the type utilized.

- Lignin is extracted from the source either physically, chemically, or biologically and varies in quality depending on the technology.

- The greatest majority of industrial lignins are chemically extracted from their native sources.

- Lignin described through their bond type or content of S-, P-, and C- type alcohols, due to their complexity, contains various chemical groups interesting for a wide array of industries.

- Lignin cannot be utilized in its original form; due to its heterogeneity and impurities from the treatment of the source material, however, it can be valorized through fractionation to enhance its homogeneity.

- Hydrogenolysis, pyrolysis, HTL, and HTC are examples of lignin valorization processes.

- There are three synthesis techniques to produce lignin-based epoxy resins. They are as follows: (a) simple blending of lignin and petroleum-based epoxy, (b) epoxidation of modified lignin, and (c) epoxidation of unmodified lignin.

- In the simple blending approach, lignin can replace a small percentage of petroleum-based BPA. In contrast, the other two techniques can achieve complete substitution of BPA.

Author details

Mohamed H. El-Moayed[1*], Justin Kühn[2], Sea-Hyun Lee[2], Mahmoud Farag[3] and Sherif Mehanny[1]

1 Faculty of Engineering, Mechanical Design and Production Department, Cairo University, Giza, Egypt

2 Institute of Textile Technology, RWTH Aachen University, Aachen, Germany

3 Department of Mechanical Engineering, School of Sciences and Engineering, The American University in Cairo, New Cairo, Cairo, Egypt

*Address all correspondence to: mohamed.elmoayed@eng.cu.edu.eg

References

[1] Mehanny S et al. Extraction and characterization of nanocellulose from three types of palm residues. Journal of Materials Research and Technology. 2021;**10**:526-537. DOI: 10.1016/j.jmrt.2020.12.027

[2] Mehanny S, Ibrahim H, Darwish L, Farag M, El-Habbak A-HM, El-Kashif E. Effect of Environmental Conditions on Date Palm Fiber Composites, in Date Palm Fiber Composites. Singapore: Springer Singapore; 2020. pp. 287-320

[3] Refate A, et al. Influence of electrospinning parameters on biopolymers nanofibers, with emphasis on cellulose & chitosan. Heliyon. Cell Press [Under review]

[4] Mehanny S, Ammar Z, Adly M. Influence of natural fiber content on the frictional material of brake pads – A review. Journal of Polymers and the Environment. Springer [Under review]

[5] Mehanny S, Magd EEA-E, Sorbara S, Navarro J, Gil-San-Millan R. Spanish poplar biomass as a precursor for nanocellulose extraction. Applied Sciences. 2021;**11**(15):6863. DOI: 10.3390/app11156863

[6] Mehanny S, Farag M, Rashad RM, Elsayed H. Fabrication and characterization of starch based bagasse fiber composite. In: Design, Materials and Manufacturing, Parts A, B, and C. Vol. 3. ASME. 2012. pp. 1345-1353. DOI: 10.1115/IMECE2012-86265

[7] Brennicke A, Schopfer P. Pflanzenphysiologie. Heidelberg: Spektrum Akademischer Verlag; 2010

[8] Roux C, Robertson J. FIBERS — Types. In: Encyclopedia of Forensic Sciences. Elsevier; 2000. pp. 838-854. DOI: 10.1006/rwfs.2000.0513

[9] Lourenço A, Pereira H. Compositional variability of lignin in biomass. In: Lignin - Trends and Applications. London: InTech; 2018

[10] Krüger G. Lignin - seine Bedeutung und Biogenese. Chemie unserer Zeit. 1976;**10**(1):21-29. DOI: 10.1002/ciuz.19760100104

[11] Zhang Y, Naebe M. Lignin: A review on structure, properties, and applications as a light-colored UV absorber. ACS Sustainable Chemistry & Engineering. 2021;**9**(4):1427-1442. DOI: 10.1021/acssuschemeng.0c06998

[12] Espinoza-Acosta JL, Torres-Chávez PI, Olmedo-Martínez JL, Vega-Rios A, Flores-Gallardo S, Zaragoza-Contreras EA. Lignin in storage and renewable energy applications: A review. Journal of Energy Chemistry. 2018;**27**(5):1422-1438. DOI: 10.1016/j.jechem.2018.02.015

[13] Fan L et al. Fast microwave-assisted catalytic co-pyrolysis of lignin and low-density polyethylene with HZSM-5 and MgO for improved bio-oil yield and quality. Bioresource Technology. 2017;**225**:199-205. DOI: 10.1016/j.biortech.2016.11.072

[14] Cao L et al. Hydrothermal liquefaction of agricultural and forestry wastes: State-of-the-art review and future prospects. Bioresource Technology. 2017;**245**:1184-1193. DOI: 10.1016/j.biortech.2017.08.196

[15] Mohamed GM, El-Shafey OI, Fathy NA. Preparation of carbonaceous hydrochar adsorbents from cellulose and

lignin derived from rice straw. Egyptian Journal of Chemistry. 2017;**60**(5):793-804. DOI: 10.21608/ejchem.2017.1311.1080

[16] Chandrasekaran SR et al. Antioxidants from slow pyrolysis bio-oil of birch wood: Application for biodiesel and biobased lubricants. ACS Sustainable Chemistry & Engineering. 2016;**4**(3):1414-1421. DOI: 10.1021/acssuschemeng.5b01302

[17] Biswas B, Singh R, Kumar J, Khan AA, Krishna BB, Bhaskar T. Slow pyrolysis of prot, alkali and dealkaline lignins for production of chemicals. Bioresource Technology. 2016;**213**:319-326. DOI: 10.1016/j.biortech.2016.01.131

[18] Guo D, Wu S, Lyu G, Guo H. Effect of molecular weight on the pyrolysis characteristics of alkali lignin. Fuel. 2017;**193**:45-53. DOI: 10.1016/j.fuel.2016.12.042

[19] Ciliberti C et al. Syngas derived from lignocellulosic biomass gasification as an alternative resource for innovative bioprocesses. Processes. 2020;**8**(12):1567. DOI: 10.3390/pr8121567

[20] Han T, Yang W, Jönsson PG. Pyrolysis and subsequent steam gasification of metal dry impregnated lignin for the production of H2-rich syngas and magnetic activated carbon. Chemical Engineering Journal. 2020;**394**:124902. DOI: 10.1016/j.cej.2020.124902

[21] Rahimi A, Ulbrich A, Coon JJ, Stahl SS. Formic-acid-induced depolymerization of oxidized lignin to aromatics. Nature. 2014;**515**(7526):249-252. DOI: 10.1038/nature13867

[22] Mau V, Gross A. Energy conversion and gas emissions from production and combustion of poultry-litter-derived hydrochar and biochar. Applied Energy. 2018;**213**:510-519. DOI: 10.1016/j.apenergy.2017.11.033

[23] Kang S, Li X, Fan J, Chang J. Hydrothermal conversion of lignin: A review. Renewable and Sustainable Energy Reviews. 2013;**27**:546-558. DOI: 10.1016/j.rser.2013.07.013

[24] Nizamuddin S et al. An overview of effect of process parameters on hydrothermal carbonization of biomass. Renewable and Sustainable Energy Reviews. 2017;**73**:1289-1299. DOI: 10.1016/j.rser.2016.12.122

[25] Oldring PKT. In: Meyers RA, editor. Coatings, Colorants, and Paints, in Encyclopedia of Physical Science and Technology. 3rd ed. Elsevier; 2003. pp. 175-190

[26] Kangishwar S, Radhika N, Sheik AA, Chavali A, Hariharan S. A Comprehensive Review on Polymer Matrix Composites: Material Selection, Fabrication, and Application, no. 0123456789. Berlin Heidelberg: Springer; 2022

[27] Kausar A. Polyurethane/Epoxy Interpenetrating Polymer Network, in Aspects of Polyurethanes. London, UK, London: InTech; 2017. pp. 1-16

[28] Zhen X et al. Facile synthesis of lignin-based epoxy resins with excellent thermal-mechanical performance. International Journal of Biological Macromolecules. 2021;**182**:276-285. DOI: 10.1016/j.ijbiomac.2021.03.203

[29] Czub P, Sienkiewicz A. In: Parameswaranpillai J, Rangappa S, Siengchin S, Jose S, editors. Synthesis of Bio-Based Epoxy Resins, in Bio-Based Epoxy Polymers, Blends and Composites. Wiley; 2021. pp. 1-72

[30] Du L, Wang Z, Li S, Song W, Lin W. A comparison of monomeric phenols produced from lignin by fast pyrolysis and hydrothermal conversions. International Journal of Chemical

Reactor Engineering. 2013;**11**(1):135-145. DOI: 10.1515/ijcre-2012-0085

[31] Xu C, Ferdosian F. Lignin-Based Epoxy Resins, in Conversion of Lignin into Bio-Based Chemicals and Materials. Berlin Heidelberg: Springer; 2017. pp. 111-131

[32] Nikafshar S, Wang J, Dunne K, Sangthonganotai P, Nejad M. Choosing the right lignin to fully replace bisphenol A in epoxy resin formulation. ChemSusChem. 2021;**14**(4):1184-1195. DOI: 10.1002/cssc.202002729

[33] Simionescu CI et al. Lignin/epoxy composites. Composites Science and Technology. 1993;**48**(1-4):317-323. DOI: 10.1016/0266-3538(93)90149-B

[34] Behin J, Rajabi L, Etesami H, Nikafshar S. Enhancing mechanical properties of epoxy resin using waste lignin and salicylate alumoxane nanoparticles. Korean Journal of Chemical Engineering. 2018;**35**(2):602-612. DOI: 10.1007/s11814-017-0301-0

[35] Pan H, Sun G, Zhao T, Wang G. Thermal properties of epoxy resins crosslinked by an aminated lignin. Polymer Engineering and Science. Apr 2015;**55**(4):924-932. DOI: 10.1002/pen.23960

[36] Zhang Y et al. Preparation and characterization of chemical grouting derived from lignin epoxy resin. European Polymer Journal. 2019;**118**(May):290-305. DOI: 10.1016/j.eurpolymj.2019.05.003

[37] Berlin A, Balakshin M. Industrial Lignins, in Bioenergy Research: Advances and Applications, no. Elsevier; 2014. pp. 315-336. DOI: 10.1016/B978-0-444-59561-4.00018-8

[38] Mandlekar N et al. An overview on the use of lignin and its derivatives in fire retardant polymer systems. In: Lignin - Trends and Applications. London, UK, London: InTech; 2018. p. 13

Section 2

Application of Lignin

Chapter 4

Emulsion Stabilization with Lignosulfonates

Jost Ruwoldt

Abstract

Lignosulfonates are biobased surfactants and specialty chemicals. Due to their amphiphilic nature, they can be utilized in many technical applications, such as plasticizers, dispersants, stabilizers, and agrochemical formulations. Here, their ability to stabilize emulsions plays an important role. This chapter hence explains the fundamentals of emulsion stabilization with lignosulfonates. First, basic concepts are introduced along with the production and chemical make-up of lignosulfonates. Second, the interfacial activity is discussed. Parameters that affect interfacial activity and emulsion stabilization efficiency of lignosulfonates are furthermore treated. Such parameters may include salinity, pH, the presence of cosolvents and cosurfactants. Third, the underlying mechanisms of emulsion stabilization are outlined. The goal of this chapter is to introduce the reader to important fundamentals, and to draw the link between basic research and industrial practice.

Keywords: lignin, lignosulfonates, surfactant, emulsion, dispersion, stabilizer, interfacial activity, colloid chemistry

1. Introduction

Lignosulfonates are biobased specialty chemicals, which are generated as a by-product during sulfite pulping of wood [1]. They are water-soluble polyelectrolytes, as they contain ionizable moieties such as sulfonate and carboxylic acid groups. Having preserved the polyaromatic backbone of lignin, they are functional amphiphiles with surfactant-like behavior. Due to their rich chemistry, lignosulfonates are found in many industrial applications [2]. This book chapter discusses the fundamentals of one of those applications, that is, emulsion stabilization. By explaining the working principles of interfacial adsorption and emulsion stabilization, the goal is to provide a reference for applied and fundamental research, as well as industry and potential end-users of lignosulfonates.

Next to cellulose, lignin is considered the second most abundant and important polymeric compound in plants [1]. Here, it fulfills functions such as water-proofing the cell-walls and adding to the mechanical cohesion and strength. While wood may contain around 20–30% lignin, its abundance in aquatic and herbaceous angiosperms is generally less. In contrast to α-cellulose, lignin is an amorphous material with a randomly branched structure. This non-uniformity can make technical utilization challenging, as reactivity and availability of functional groups may vary [3]. Still, some applications may benefit from polydispersity. For specialty chemicals such as surfactants, it is the function or the effect that counts. Salinity changes may induce precipitation of an ionic surfactant. If the surfactant is polydisperse, precipitation may be limited to only a fraction of it. The remaining surfactant can then still perform the desired function. Moreover, mixtures of dissimilar surfactants can form synergies, where the performance may be superior over its individual components [4]. The application of lignin as specialty chemicals can thus benefit from its inherent characteristics, which includes its polydispersity. It is therefore not surprising that dispersants and surfactants are currently the dominant value-added applications of technical lignin.

It is frequently stated 50–70 million tons of technical lignin are produced per year, of which only 1–2% of are utilized for value-added products [5]. The majority is burned for heat and chemicals recovery in Kraft mills [6]. More than 1 million tons of lignosulfonates are produced annually [7], hence dominating the market of technical lignin. In contrast to many other lignin products, lignosulfonates are readily water-soluble. Their most common application is dispersants [8], for example as concrete plasticizers, drilling mud thinners, coal-slurry or dye dispersants [9]. Roughly 50% of the annual production of lignosulfonates were used for concrete admixtures in 1999 [10]. Other applications include dust binders, chelating and complexing agents, soil conditioning agents, floatation agents, and water-in-oil emulsion stabilizers [8, 9, 11, 12]. As the effect on interfacial tension is limited, high shear is usually required during emulsification [13]. Still, lignosulfonates can yield emulsions that are stable over months and longer, finding application for example in agrochemical formulations [8]. Other potential areas with less commercial importance include corrosion and scale inhibitors, CO_2 flooding and enhanced oil recovery, as well as polymer precursors and additives [2, 14–20].

Great importance is attributed to biopolymers these days, as they are inherently renewable and largely biodegradable. By substituting fossil-based polymers with biopolymers, the transition to more sustainable technologies is furthered. Expanding the use of lignosulfonates is therefore an important piece in the puzzle. Compared to the global production of surfactants or pulp and paper products, lignosulfonates are currently a niche application. There is hence a potential for growth, in particular for applications that involve liquid–liquid interfaces.

2. Fundamentals

2.1 Definitions and distinctions

The term **lignin** is used for the natural lignin as found in lignocellulosic biomass. It is also referred to as pristine lignin, as no treatment or modification has been done. Chemically speaking, lignin is a biopolymer consisting of the monolignol units sinapyl alcohol (S), coniferyl alcohol (G), and *p*-coumaryl alcohol (H) [21], which are connected by various oxygen- and carbon–carbon linkages. Lignin has been described as a polyaromatic and randomly branched biopolymer.

Technical lignin refers to the lignin-rich product obtained from biomass separation processes. While pristine lignin is a virtually “infinite” network [22], technical lignin is a fragmentated version thereof. The molecular weight, composition, and ratio of functional groups of technical lignin are hence different. Their abundance can be affected by parameters such as biomass feedstock, separation process, and purification steps. Technical lignin usually has a purity of at least 70% (Klason lignin + acid soluble lignin), with commercial products being closer to 85–100% of lignin per ash-free dry matter.

Pulping of lignocellulose biomass is conducted to obtain a fibrous material. Approximately 90% of the global pulp products are made from wood, whereas 10% originate from annual plants [1]. Common end-products include paper, cardboard, molded pulp, and specialty cellulose. Mechanical pulping applies force in a refiner to defibrate the feedstock into fibers and fibrils. Chemical pulping dissolves the lignin and other substances to liberate the cellulose fibers. Industrial processes can also be based on a combination of mechanical and chemical treatments. Today, technical lignin most commonly originates from chemical pulping.

A **biorefinery** is defined as "the sustainable processing of biomass into a spectrum of marketable products (food, feed, materials, chemicals) and energy (fuels, power, heat)" [23]. In other words, a biorefinery separates biomass into useful materials, chemicals, and energy. First generation biorefineries differ from second generation biorefineries by the choice of input material, i.e., food crops (first generation) instead of non-food crops (second generation) [24]. Pulping of wood is thus defined as a lignocellulosic crop, second generation biorefinery.

An overview of existing **biomass separation processes** with the resulting lignin product is given in **Figure 1**. Technical lignin has traditionally been obtained by chemical pulping, i.e., Kraft, sulfite, or soda pulping. Organosolv pulping is a more recent invention, which utilizes organic solvents and often acid catalysis for delignification [26]. The original goal of organosolv pulping was to obtain a fiber material. Still, this process has also attracted interest for biorefinery applications, which subsequently process the cellulose for chemical utilization [27]. The production of technical lignin can hence be conducted as part of a pulping processes, but it is not limited to that. An example for a non-pulping type lignin would be hydrolysis lignin, which is produced in biorefinery sugar platforms. Here, the cellulose and hemicellulose are hydrolyzed to extract sugar monomers and oligomers. The sugar can further be converted to higher value products, for example ethanol or furfural. The residual solids are rich in lignin and therefore termed hydrolysis lignin. Residual cellulose and other impurities can impart poorer performance and hence lower value on this lignin type [3, 28]. Still, recent developments have yielded increased purity and reactivity, e.g., the Cellunolix® lignin by St1 (Finnland) [29] or the Lignova™ lignin by Fibenol (Estonia) [30]. Because of these developments, it can be expected that hydrolysis lignin may play a larger role in the future. Steam explosion lignin is viewed as a separate category to hydrolysis lignin by some authors [3, 31]. Yet, both lignin types

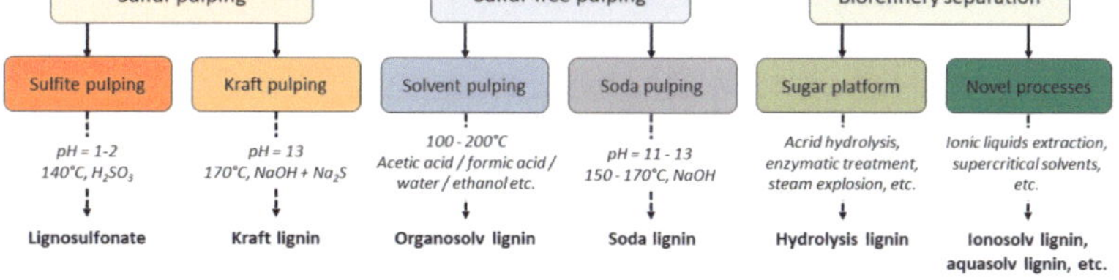

Figure 1.
Lignin extraction processes and their dominant products. Modified from Laurichesse and Avérous [25].

are similar in composition and the treatments are sometimes even cascaded. Steam explosion treatment subjects the substrate to steam at elevated temperature (ca. 160–280°C) and pressure (ca. 7–48 bar) [32]. This induces biomass disintegration and partial hydrolysis. Due to this, it will be considered as part of hydrolysis lignin, as shown in **Figure 1**. Hydrolysis lignins can further be subdivided into acid hydrolysis lignin and enzymatic hydrolysis lignin, with the latter being more reactive due to a lower degree of condensation [3]. Recent advances have also yielded novel processes, which are still in an early stage of their development. Ionic liquids have been demonstrated to function as lignin solvents [33]. Combined with other treatments, ionic liquids can yield new products and conversion routes with promising features; however, some economic and technical challenges still need to be addressed [34]. Another recent invention is the use of supercritical solvents, e.g., supercritical water yielding products such as Aquasolv lignin [35].

Lignosulfonates are traditionally extracted from sulfite spent liquors. Sulfite pulping is usually operated at low pH, but neutral or alkaline sulfite pulping have also been developed [1]. Sulfite liquors typically contain about 50–80% lignosulfonates, 30% hemicellulose and 10% inorganics per dry matter content [5]. Purification is often conducted by membrane filtration, removing low molecular weight components such as sugars. Commercial grade lignosulfonates are available in both purified and unpurified qualities. As a result of the pulping conditions, they also tend to exhibit a higher degree of condensation than, e.g., soda or organosolv lignin [3].

Sulfonated lignin is produced by chemical modification of lignin separated by a process other than sulfite pulping [2]. For example, MeadWestvaco produces Kraft lignin sulfonated with sulfite salts and an aldehyde, e.g., formaldehyde [5]. A variety of modification processes exist and the differences between lignosulfonates and sulfonated lignin can be marginal.

The **counterion** is the ion accompanying a second ionic species to maintain charge neutrality. Lignosulfonates contain covalently bond anionic moieties, which most importantly include sulfonate and carboxyl groups. A counter ion with positive charge is hence necessary for charge neutrality. This counterion is frequently a remnant of the pulping process, i.e., pulping with sodium, calcium or magnesium bisulfite will yield sodium, calcium, or magnesium lignosulfonates, respectively. Ion-exchange may also be conducted to replace the counterion. Protonation is furthermore possible, i.e., hydrogen as the counter ion. However, lignosulfonate dispersants are usually the salt of lignosulfonic acid, as this yields a more moderate pH and improves water-solubility.

A **surfactant**, i.e., surface-active agent, is a compound that can lower the surface or interfacial tension of a liquid in contact with another phase [36]. This effect is usually accompanied by adsorption at the surface or interface, i.e., enrichment at the phase boundary [37]. Surfactants usually contain hydrophilic and lipophilic moieties, which facilitate interfacial adsorption. Lignosulfonates have been shown to reduce the surface tension of water [38] and can hence be classified as surfactants.

Surface tension can be observed as the tendency of a liquid surface to assume the smallest possible surface area. It is linked to the intermolecular attraction forces within the liquid and is commonly denoted as force per unit length or energy per unit area. Dispersing a liquid with high surface tension hence requires more energy than dispersing a liquid with low surface tension. In this chapter, the surface tension will be used when discussing liquid–gas phase boundaries.

The **interfacial tension** is the equivalent to the surface tension at liquid–liquid interfaces. A system with low interfacial tension requires less work for emulsification than a system with high interfacial tension.

2.2 Production

Lignosulfonates are produced as a by-product during sulfite pulping of wood [1]. Sulfite pulping is a long-established method for producing cellulosic fibers [39]. The process usually operates at low pH, utilizing sulfite or bisulfite salts to soften and remove the lignin. The lignin undergoes a number of reactions, which include [1]:

- Hydrolysis
- Sulfonation
- Sulfitolysis
- Dissolution
- Degradation
- Condensation

During hydrolysis, the lignin-carbohydrate and, to a smaller extent, lignin-lignin linkages are broken down [1]. Sulfonation introduces sulfonate groups onto the lignin. When present in sufficient amounts, the sulfonate groups render the lignin water-soluble, which facilitates dissolution in the cooking liquor. Sulfitolysis and degradation may further reduce the molecular weight, whereas condensation reactions increase it. Condensation counteracts delignification by forming new carbon–carbon bonds [1].

Natural lignin is synthesized from the three monolignols, i.e., sinapyl alcohol (S), coniferyl alcohol (G), and p-coumaryl alcohol (H) units [21]. Coniferous lignin (softwood lignin) is composed mainly of G-units, whereas lignin from broad-leaved trees (hardwood lignin) contains a mixture of G- and S-units [40]. Lignin from annual plants also contains H-units in addition to G- and S-units [41]. The feedstock therefore affects the composition of the resulting lignosulfonates, as the three monolignols vary in their methoxy content and potential branching. The feedstock can furthermore affect the pulping process, since sulfonation of hardwood lignin is slower than of softwood lignin [1]. Moreover, softwood lignin is more prone to condensation reactions, resulting in higher molecular weight as compared to hardwood lignosulfonates [42]. An overview of the monolignols and a schematic of softwood lignin is given in **Figure 2**.

While alkali lignin can be separated by acid precipitation, this approach is not feasible for lignosulfonates. Instead, membrane filtration (ultrafiltration) is often conducted to purify the sulfite liquor [5]. Challenges can arise due to overlapping molecular weight of the lignosulfonates and dissolved hemicellulose. One approach is to cascade a series of membranes with different cut-off molecular weights [44]. An alternative is given by the Howard process [5], which uses lime (calcium oxide) to precipitate the lignosulfonate above pH 12 [45]. The precipitated solids can be separated mechanically and washed to improve the purity. Other potential approaches include amine extraction, electrolysis, ion-exchange resins, the Pekilo process (fermentation and ultrafiltration), and reverse osmosis [5].

Several factors affect the composition of lignosulfonates, which include but are not limited to:

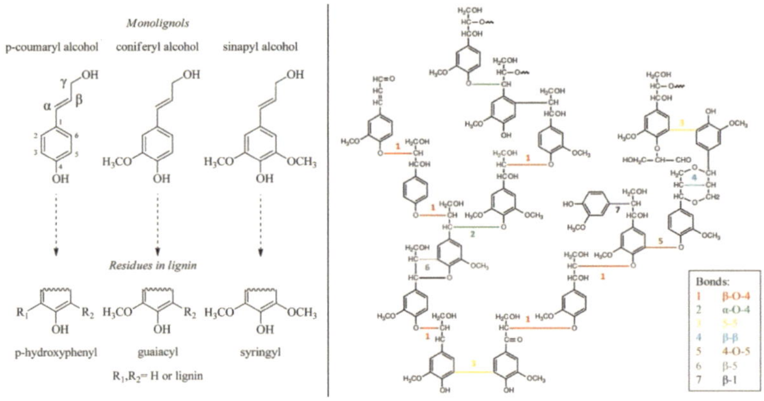

Figure 2.
Primary lignin monomers and their corresponding units (left) [25]. Schematic structure of softwood lignin (right) [43].

- The choice of raw material, e.g., softwood or hardwood, affecting the monolignol composition
- The choice of pulping chemicals
- The pulping conditions
- Recovery and purification
- Post-treatment and chemical modification

The composition of lignosulfonates is inherently linked to its characteristics and behavior, and hence to the performance in technical applications [2].

2.3 Structure and composition

As discussed in the previous chapter, the structure and composition of lignosulfonates is strongly dependent on their origin and production. Still, there are certain characteristics that are worth discussing.

Due to their structure and monomeric configuration, cellulose and hemicellulose contain a higher percentage of oxygen than lignin. Lignin furthermore has the highest carbon content of these three biopolymers. A unique feature of lignosulfonates is their elevated sulfur content, which arises from the pulping process. **Table 1** lists a range of elementary composition, as published in current literature. It should be mentioned that these values should only be taken as indicators, since individual samples may be different. In particular, the ash content may exhibit values up to 40 wt.%. Still, an elevated oxygen content and lower carbon content can in theory indicate lower purity, as the contributions from cellulose and hemicellulose would be greater.

The chemical structure of lignosulfonates mirrors that of both native lignin and the modifications done during sulfite pulping. As such, the skeletal configuration of lignin is preserved to some extent, while new carbon–carbon linkages have been formed. In addition, sulfonate and carboxylic acid groups were added, which are not found in native lignin to this extent. Generic structure models of lignosulfonates are shown in **Figure 3**.

Parameter	Approximate content in wt.%	Reference
Dry matter	94–96	[46]
Ash	4–10	[3, 47]
Carbon	29–47	[48–50]
Hydrogen	5	[48, 50]
Oxygen	37–54	[48–50]
Sulfur	2–10	[46]
Nitrogen	0.02	[3]

Table 1.
Approximate elementary composition of lignosulfonates.

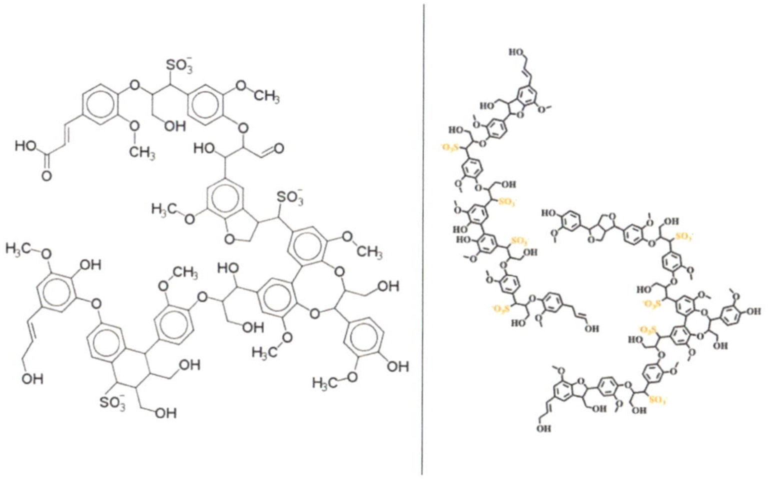

Figure 3.
Generic (simplified) structure of lignosulfonates according to Kun and Pukanszky [51] (left), and Fiorani et al. [52] (right).

It is important to note that these are only approximate models, as lignosulfonates are a polydisperse mixture of many different macromolecules. The structure hence varies not only between different lignosulfonates, but also within a given sample. In other words, lignosulfonates should be considered statistical entities rather than classical chemical compounds [53].

The chemical moieties of lignosulfonates can be categorized into ionizable, polar and non-polar groups [2]. Ionizable groups include sulfonate, carboxylic acid, and phenolic hydroxyl groups. At a neutral pH of 7, mainly the first two can be considered dissociated (ionized) and hence hydrophilic. Phenolic hydroxyl groups are usually ionized at around pH 9–10 [54, 55], but values extending to pH 5–12 have been reported [56]. Polar groups include various oxygen containing moieties due to the higher dipole moment of oxygen–carbon and oxygen-hydrogen linkages. These include ketones, aldehydes, and methoxy groups. Despite being polar, these groups are not considered operative hydrophilic groups [57]. Aliphatic hydroxyl and ether groups are also present to a great extent in lignosulfonates. These can be intrinsically hydrophilic; however, their functionality is determined by the surrounding molecular structure [57–59]. Because of this, the water-solubility of lignosulfonates arises from the presence of ionizable groups, i.e., sulfonate and carboxyl groups at neutral pH. Nonpolar groups include aromatic and aliphatic units, as found in the skeletal configuration of lignin. An overview of the common functional groups and linkages is given in **Figure 4**. This overview does not include additional functionalities, which may be grafted onto the lignosulfonates, e.g., by phosphorylation, alkylation, sulfobutylation or silylation [60].

The abundance of functional groups can be of interest for two reasons. Firstly, this is an important parameter for chemical modification, as specific functional groups may be targeted. Secondly, the physicochemical properties of lignosulfonates are highly dependent on their composition. Hydrophobic interaction chromatography (HIC) has played an important role in characterizing lignosulfonates recently [61–63]. Based on this technique, the charge-to-size ratio was reportedly lower for more hydrophobic lignosulfonates [64]. Better performance as suspension or emulsion stabilizer was furthermore seen for more hydrophobic lignosulfonates [13, 63]. This example illustrates, how the abundance of functional groups may impact the performance in technical applications.

Compared to other technical lignins, lignosulfonates tend to exhibit a lower amount of phenolic hydroxyl groups [47]. While the sulfonate group is a distinct feature of lignosulfonates, the abundance of carboxylic acid groups is comparable to that of other technical lignins. A recent study was performed aqueous carbon black dispersions stabilized by sodium lignosulfonate, which also listed the composition of six commercial samples [46]. The values are summarized in **Table 2**. As all samples

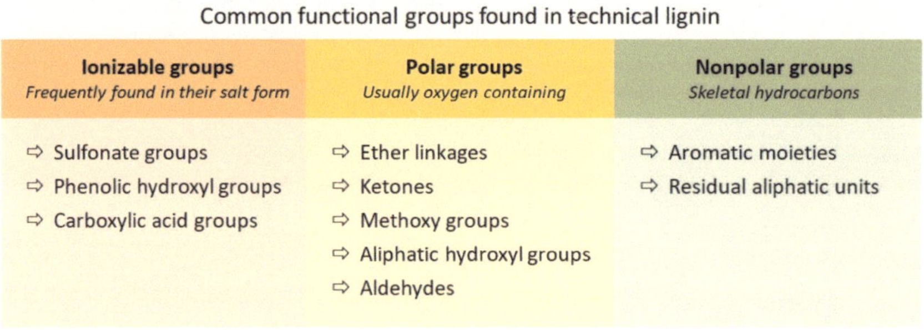

Figure 4.
Commonly encountered chemical bounds and functional groups in technical lignin. Image taken from [2].

Parameter	Approximate range
M_w in g/mol	1800–4000
M_n in g/mol	6000–330,000
Polydispersity index	3–83
Organic sulfur (wt.%)	2.1–9.4
Carboxyl groups (wt.%)	8.3–14.9
Phenolic hydroxyl groups (wt.%)	1.3–2.0
Methoxy groups (wt.%)	7.3–15.4

Table 2.
Abundance of functional groups of softwood sodium lignosulfonates according to Subramanian et al. [46].

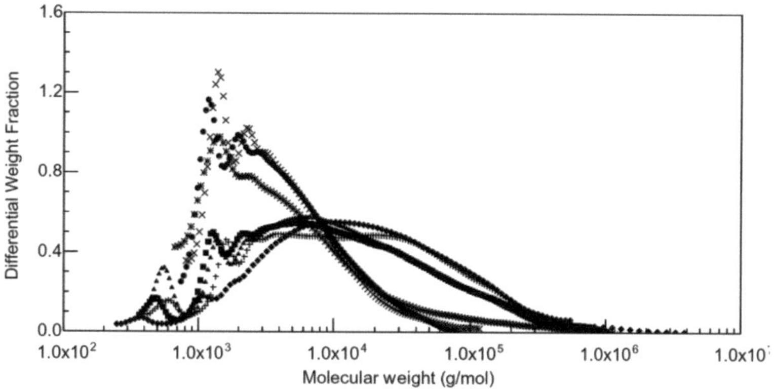

Figure 5.
Molecular weight distribution of lignosulfonates originating from spruce [LS1 (♦), LS2 (■) and LS3 (+)], a spruce-birch blend [LS5 (▲)], aspen [LS6 (*)], and eucalyptus [LS7 (●) and LS8 (×)]. Image taken from [42].

were of softwood origin, a higher methoxy content can be expected for hardwood lignosulfonates due to the monolignol composition.

Table 2 also lists the number-average (M_n) and mass-average molecular weight (M_w). A polydispersity index of up to 83 accounts for a great variety in molecular mass, which is characteristic for lignin from sulfite pulping. The molecular weight can have several implications on the properties of lignosulfonates. It is naturally linked to the diffusion coefficient [65], which will further affect interfacial adsorption and related phenomena [13, 66]. Research has shown that the degree of sulfonation decreases with increasing molecular weight [67]. Stearic screening can occur in lignosulfonates with high molecular weight [3]. This effect was used to explain, e.g., the lower effect on zeta potential by shielding ionizable groups [68] or the lower reactivity by screening of phenolic hydroxyl groups [3]. Overall, the molecular weight of lignosulfonates may span from less than 1000 g/mol to more than 400,000 g/mol [42, 67]. The molecular weight ranges are thus higher than for other technical lignins [2]. In comparison, values of (1000–15,000 g/mol) have been reported for soda lignin, 1500–25,000 g/mol for Kraft lignin, and 500–5000 g/mol for organosolv lignin [3]. Acid hydrolysis lignin is closer to lignosulfonates in terms of molecular weight, where values of 1500–50,000 g/mol have been reported [69]. The molecular weight distribution of various lignosulfonates is illustrated in **Figure 5**. As can be seen, the molecular weight of the hardwood samples was consistently lower than that of softwood lignosulfonates. This difference is likely the result of the monolignol configuration, which can further affect the sulfite pulping process.

3. Adsorption on surfaces and interfaces

3.1 Adsorption

It is generally agreed that adsorption of lignosulfonates follows the Langmuir isotherm [70–73]. The according framework was first published by Irving Langmuir in 1916, which described the adsorption of gasses on solids [74]. A common expression of the Langmuir equation is given in Eq. (1), were Γ corresponds to the amount of adsorbed species, Γ_{max} is the maximum amount adsorbed, c is the concentration of adsorbing species in bulk, and K is the Langmuir equilibrium constant.

$$\Gamma = \frac{\Gamma_{max} Kc}{1 + Kc} \tag{1}$$

The parameters Γ_{max} and K can be determined via mass-balancing, i.e., measuring the adsorbed amount in dependence of concentration. This approach is simple for adsorption on solid surfaces, as the amount of non-adsorbing surfactant is equivalent to the remaining bulk concentration. An example for the adsorption of lignosulfonates on carbon black is given in **Figure 6**. The characteristic behavior accompanying Langmuir adsorption isotherms is a sharp increase of adsorbed amount at low concentrations, whereas higher concentrations yield a plateau.

Determination of the Langmuir equation parameters to liquid–liquid systems is less straight forward than for liquid–solid dispersions. Emulsification is accompanied by an increase in specific surface area, which can be difficult to quantify. The surface or interfacial tension can be used instead of mass-balancing. Langmuir type adsorption implies the applicability of the Szyszkowski equation, as given in Eq. (2). Here, the surface pressure Π is calculated as the difference of surface or interfacial tension without surfactant (γ_0) and at concentration c ($\gamma(c)$). The equation further involves the temperature T, the ideal gas constant R, and the constant n.

$$\Pi = \gamma_0 - \gamma(c) = n\text{RT}\Gamma_m \ln(1 + Kc) \tag{2}$$

The surface excess Γ_m can further be determined using the Gibbs adsorption isotherm. As stated in Eq. (3), this framework relies on the linear-logarithmic progression of surface tension γ with concentration c. It is important to note that the linear-logarithmic regime does not span over the entire concentration range. At low surfactant concentrations, the surface or interfacial tension approaches the surfactant-free case γ_0 asymptotically. At high lignosulfonate concentrations, agglomeration and

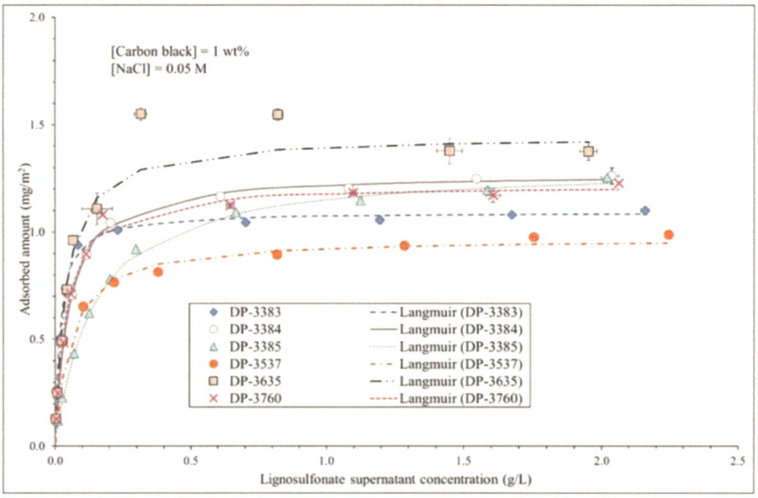

Figure 6.
Adsorption isotherm of lignosulfonate on 1 wt% carbon black along with Langmuir isotherm fitting. Image taken from [46].

other effects will decrease the observed slope [38, 75]. A more detailed discussion of the surface and interfacial tension of lignosulfonates is given in Section 3.2.

$$\Gamma_m = -\frac{1}{n\mathrm{RT}}\left(\frac{\partial \gamma}{\partial \ln c}\right) \tag{3}$$

Assumptions can furthermore be made to determine the constant n. In case of strong surfactant electrolytes, such as $R - O - SO_3^- Na^+$, both the surfactant ion and the counterion must absorb, yielding a value of $n = 2$ [76]. A value of $n = 1$ corresponds to non-ionic surfactants or ionic surfactants in presence of high concentrations of an indifferent electrolyte, such as NaCl. Both assumptions have been applied to lignosulfonates and were found to yield conclusive results [13, 66].

By applying the above framework, other information can also be extracted from surface and interfacial tension measurements. For example, the surface excess Γ_m can be used to calculate the area per molecule, as stated in Eq. (4) [76]. Here, N_{av} denotes the Avogadro's constant $6.022 \times 10^{23}\ \mathrm{mol}^{-1}$.

$$\mathrm{A_m} = \frac{1}{\Gamma_m N_{\mathrm{av}}} \tag{4}$$

Several authors have studied the adsorption kinetics of lignosulfonates on solid surfaces [71, 72, 77–80]. The reports generally agree that pseudo second-order kinetics provided the best fit. Such kinetic can be expressed as written in Eq. (5), where q_t is the adsorbed amount at time t, q_e is the amount adsorbed at equilibrium, and k is the rate constant.

$$\frac{\mathrm{t}}{\mathrm{q_t}} = \frac{1}{kq_e^2} + \frac{t}{q_e} \tag{5}$$

Eq. (5) can be fitted to experimental data by plotting $\frac{t}{q_t}$ against t, yielding $\frac{1}{q_e}$ as the slope and $\frac{1}{kq_e^2}$ as the intersection. This model is, however, limited to adsorption from liquid onto solid phase. Kinetic modeling of lignosulfonate adsorption at liquid–liquid interfaces has also been attempted; yet, the results indicated that water–oil adsorption was not diffusion-controlled [66]. It usually takes hours or days to attain an equilibrium state at the liquid-air or liquid–liquid interface [66, 81, 82]. This time span is greater than for simple monodisperse surfactants. Different explanations have been given for this behavior. The lignosulfonate can undergo diffusion exchange at the interface, replacing molecules with a lower diffusion coefficient but higher effect on interfacial tension. In addition, the individual macromolecules may be subject to rearrangement, both with respect to the interface and to each other [2]. Such realignment of conformation has been described, e.g., for petroleum asphaltenes [83, 84]. Both lignosulfonates and asphaltenes share common characteristics, e.g., both are polybranched, exhibit a tendency for self-association, and require overnight storage, as the emulsions would be less stable if processed immediately after emulsification [13, 85]. Analogies have therefore been drawn between these two species in terms of their interfacial behavior [2, 66].

3.2 Effect on surface and interfacial tension

Measuring surface tension is a useful tool, e.g., for assessing the behavior of lignosulfonates in solution or their interactions with other components [38, 86]. In addition, the interfacial tension provides a measure for the ease of emulsification [13]. The effect on surface or interfacial tension is related to the amphiphilic property of lignosulfonates, containing both hydrophilic and hydrophobic moieties. From a molecular point of view, the lignosulfonate macromolecule can attain a lower state of energy by extending its hydrophilic moieties into the water, while facing the hydrophobic parts away from it. The decrease in surface tension at increasing surfactant concentration can be explained by different mechanisms. Firstly, adsorption and desorption are occurring simultaneously, where an equilibrium is attained if the adsorption rate is equal and opposite to the desorption rate. By increasing the surfactant concentration, it is statistically more likely for a surfactant molecule to be in the right position and conformation to undergo adsorption. This would hence drive the adsorption–desorption equilibrium towards a higher surface coverage. Secondly, lignosulfonates are polyelectrolytes. By increasing the lignosulfonate concentration, the electrolyte concentration is also increased. As has been shown by small-angle X-ray scattering (SAXS), the effective surface charge of lignosulfonates decreases at increasing concentration [87]. Higher concentrations of a common ion indeed enhance effects such as counterion condensation, charge screening, and the dissociation equilibrium [88]. Water-solubility is ensured by the ionic groups of lignosulfonates, so a lower effective charge would have destabilizing effect. At a lower water-solubility, the equilibrium would hence be shifted to increased surface or interface adsorption. This effect concurs with the observations that lignosulfonates can be precipitated by salt addition (salting-out) [89] and that emulsion stabilization is highest, if the stabilizers are on the verge of precipitation [90].

Lignosulfonate adsorption at aqueous surfaces is evidenced, among others, by a decrease in surface tension [13, 38, 91]. Within a certain range, this decrease follows a linear-logarithmic progression with concentration [13, 38]. Below this range, the surface tension of the pure liquid is approached asymptotically. At high lignosulfonate concentrations, the slope of the surface tension decreases. Some authors have explained this behavior with the onset of lignosulfonate aggregation [75, 92], but other effects cannot be ruled out. Measurements of the surface tension with respect to concentration are shown in **Figure 7**.

In comparison with commercial surfactants, the effect on surface or interfacial tension is often less at the same mass concentration [75, 82, 92]. However, in comparison with other polyelectrolyte polymers, the effect on surface tension can be considerably higher [38]. This circumstance is likely related to the chemical composition and structure, which define macroscopic properties such as solubility, hydrophilic–lipophilic balance (HLB), or surface coverage.

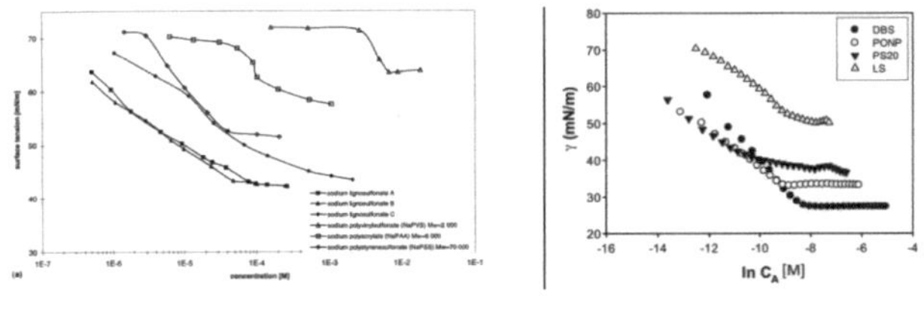

Figure 7.
Equilibrium surface tension in dependence of surfactant concentration. Comparison of lignosulfonates (LS) with polyelectrolyte polymers (left) [38] or with the surfactants dodecyl benzenesulfonate (DBS), nonylphenyl polyoxyethylene glycol (PONP) and polysorbate 20 (PS20) (right) [92].

3.3 Parameters affecting the surface activity of lignosulfonates

There are several external factors, which govern the surface activity of lignosulfonates, in addition to the intrinsic properties such as composition and abundance of functional groups. These factors include, but are not limited to:

- Salinity
- pH
- Temperature
- Co-solvents in the aqueous phase
- Oil phase composition

It has been shown that increasing salinity with simple monovalent electrolytes, such as NaCl, can have the same effect as increasing lignosulfonate concentration [38, 66]. The effect of NaCl concentration on surface or interfacial tension is illustrated in **Figure 8**, which is marked by same linear-logarithmic progression as in **Figure 7**. As discussed earlier, increasing the concentration of a common ion facilitates a lower effective surface charge, which further enhances surface or interface adsorption. The similarities in observed surface tension hence corroborate that the surface adsorption of lignosulfonates is also driven by salinity. It could further be argued that outside the tested concentration range in **Figure 8**, the same slope-decrease or plateau would be visible is in **Figure 7**. While this may be plausible, it has yet to be demonstrated experimentally.

However, the data exhibited in **Figures 7** and **8** represents a simplified case. In both instances, the counterion was solely the monovalent sodium ion, as only sodium lignosulfonate and NaCl were used. In presence of multivalent cations, the surface phenomena of lignosulfonates are more complex. Di- or trivalent cations, for example, were shown to induce interface gelling [66]. As illustrated in **Figure 9**, such interfaces behave inelastically, exhibiting wrinkles and cracks upon deformation. Pendant drop tensiometry relies on the applicability of the Young-Laplace equation. Since this is not the case for inelastic interfaces, the technique fails to predict accurate interfacial tension. Similar challenges are evident for other techniques, such as the Du Noüy ring method or spinning drop tensiometry.

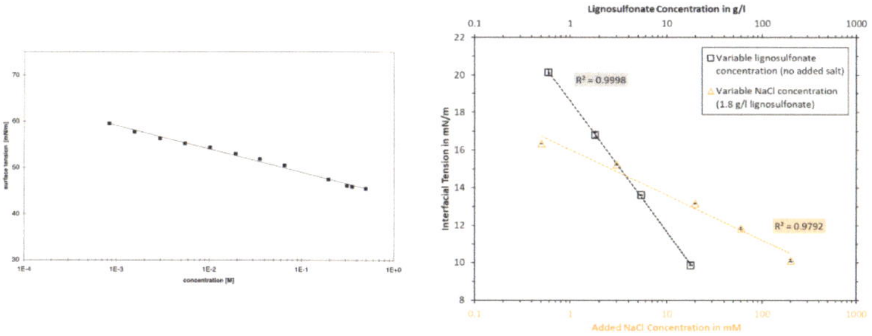

Figure 8.
Effect of NaCl concentration on the surface tension of 0.01 wt.% lignosulfonate in water (left) [38]. Effect of NaCl or lignosulfonate concentration on the tension of the water-xylene interface (right) [66].

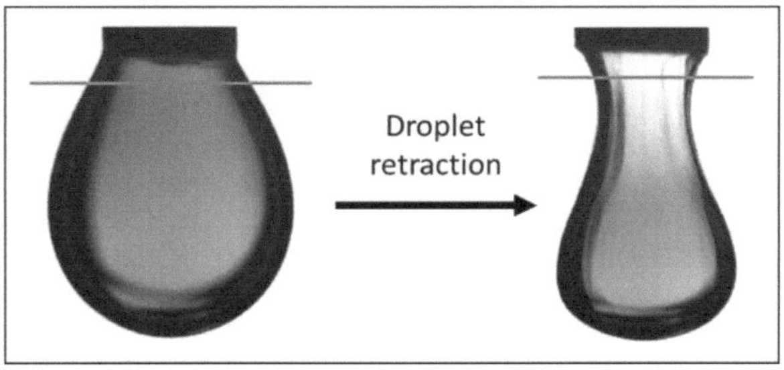

Figure 9.
Droplet retraction in pendant drop video-tensiometry. In incompressible interface layer is visible as wrinkling of the droplet surface, which was formed by sodium lignosulfonate in presence of $CaCl_2$. Image taken from [66].

Lowering the pH can further enhance the effect of lignosulfonates on surface tension [91]. This circumstance is in analogy to the salting-out effect of simple electrolytes. Phenolic moieties are said to ionize at around pH 9–10, while the carboxylic acid groups ionize at pH 3–4 [54, 55]. A lower pH can hence reduce the total charge of the lignosulfonate macromolecules. This would further reduce the water-solubility and thereby drive the equilibrium towards higher adsorption. However, the discussed effect is valid only if the lignosulfonates remain water-soluble. If precipitation is evident, then the bulk concentration would decrease, which would also reduce surface and interface adsorption.

As depicted in **Figure 10**, increasing the temperature can also increase the effect of lignosulfonates on surface tension. At higher temperature, the hydrodynamic radius of lignosulfonates decreases [93]. Thermodynamically speaking, the entropy is higher at elevated temperatures. This can enable a larger number of possible conformations, thus reducing the average molecular dimensions [88]. As a result, lignosulfonate aggregation and a reduction of zeta potential have been reported at elevated temperature [94]. These two effects indicate solution-destabilization, which could further promote surface adsorption.

Co-solvents in the aqueous phase can alter characteristics such as the solution parameter and the dielectric constant. This will inevitably also affect the surface behavior of the lignosulfonates. The addition of low molecular weight alcohols yielded a decrease of surface charge and slight interparticle association [87]. The effect was thereby similar as to increasing the ionic strength, implying solution destabilization of lignosulfonates by alcohol addition. On the other hand, hydrophobic interaction chromatography uses water/ethanol and water/2-propanol mixtures to eluate the more hydrophobic fractions [64], which would suggest that water/alcohol mixtures are a better solvent for these. This is corroborated by the fact that the surface pressure decreased due to ethanol addition [81], i.e. fewer lignosulfonate molecules would enter the interface. Corresponding measurements of interfacial tension are shown in **Figure 10**. Overall, the effects of adding alcohols are hence counteracting, i.e., reduction of effective charge (destabilizing) and solubility improvement (stabilizing). It is thus not surprising that the emulsion stability could be both improved and reduced after alcohol addition (**Figure 11**) [81].

The addition of non-solvents to water will eventually lead to lignosulfonate precipitation [87]. In this case, the interfacial activity would also decrease, as the bulk concentration is lower [66]. Solvent shifting can be useful for the production of functional micro- and nanoparticles from lignin [95]. However, the technology bears limited importance for this chapter, as the resulting Pickering-emulsions tend to be less stable [66].

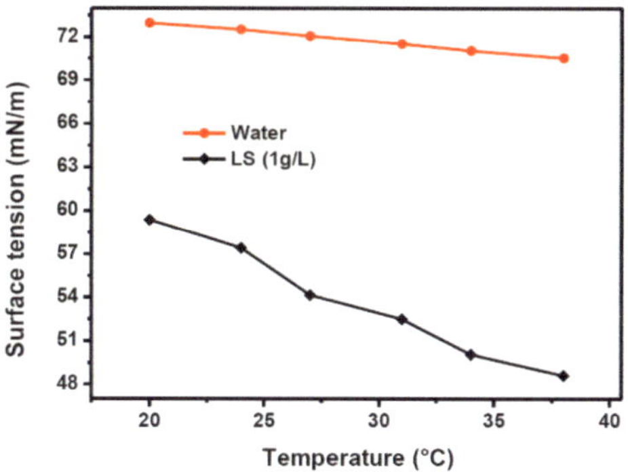

Figure 10.
Surface tension of water and 1 g/l lignosulfonate in water. Image taken from [93].

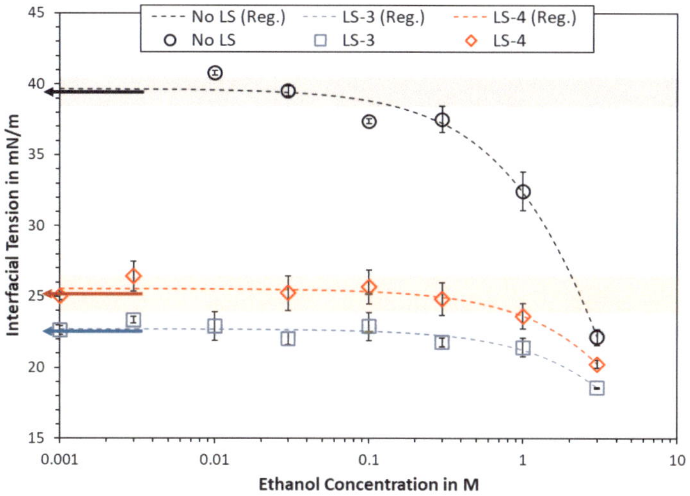

Figure 11.
Tension of the mineral oil–water interface in dependence of ethanol concentration. The samples contained 1 g/l lignosulfonate (LS) and 20 mM NaCl as background electrolyte. Figure taken from [81].

At last, the oil phase composition can also affect the interfacial activity of lignin. A simple explanation for this can be based on the concept of the hydrophilic–lipophilic balance (HLB). Surfactants can generally be classified according to their HLB, where lower HLB values account for more lipophilic surfactants and high values for more hydrophilic ones. It is generally agreed that an HLB of 3–6 is characteristic for water-in-oil (W/O) emulsifiers, whereas 8–18 are suited for oil-in-water (O/W) emulsifiers [76]. According to the HLB model, optimum emulsion stability is given, when a surfactant-blend matches the HLB value of the emulsified liquids. In analogy to that, the interfacial tension is also said to exhibit a minimum at the optimal HLB of the surfactant mixture. The concept is illustrated in **Figure 12**.

Oils with low polarity, such as paraffins and mineral oil, tend to have a required HLF for O/W emulsions of 9–11, whereas more polar oils are in the range of 12–17, e.g., oleic acid, chlorinated paraffins, and aromatics such as toluene [96]. The HLB scale can hence also be an indicator for the compatibility or affinity of surfactants towards a certain oil phase. Lignosulfonates were reported to have an HLB of 11.6 based on their composition [97], but the effective HLB is likely higher than this, as the

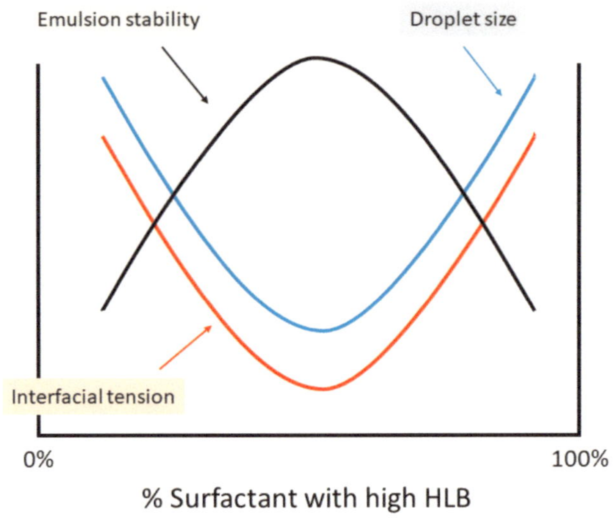

Figure 12.
Variation of emulsion stability, droplet size, and interfacial tension with percentage of surfactant with high HLB. Figure reproduced from [76].

cited value is a rough estimation based on the elementary composition. Among others, it does not consider stearic effects. In addition, the contribution of the sulfonate group to the effective HLB is generally high. There are few studies comparing the stability of different oils emulsified with lignosulfonates. Experience shows that aromatic solvents, e.g., toluene or xylene (HLB = 14–15), tend yield more stable than emulsions with mineral oil (HLB = 10) [13]. Accordingly, it was demonstrated that the effect on interfacial tension of xylene was greater than on mineral oil, as plotted in **Figure 12**. It would only make sense that lignosulfonates, comprising a polyaromatic structure, also have a higher affinity to aromatic oils than to paraffinic ones. Based on their behavior, lignosulfonates can be classified as O/W emulsion stabilizers with an effective HLB of 10–18 (**Figure 13**).

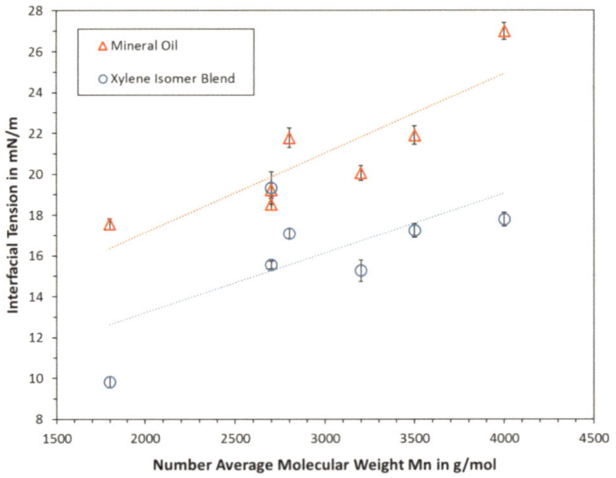

Figure 13.
Effect of molecular weight and oil phase on interfacial tension. Figure taken from [13].

3.4 Interactions with other surfactants

The presence of other surfactants can affect the surface activity of lignosulfonate. Surfactants can be distinguished based on their charge, i.e., non-ionic, anionic, cationic, and zwitterionic surfactants. Another classification would be based on the structure, that is, polymeric and non-polymeric surfactants. Electrostatic interactions with lignosulfonates are eminent, if the co-surfactant carries a charge. Surfactant interactions will hence be discussed for each individual type in this sub-chapter.

Interactions with non-ionic surfactants can be due to effects such as hydrogen bonding or hydrophobic interactions. Straight-chain alcohols can exhibit surfactant-like properties, if the chain length is at least four carbon atoms or more [98]. The effect of these was studied by Qiu et al., who concluded that alcohols can improve the surface activity of lignosulfonate [86]. The authors attributed the largest effect to alcohols with a chain-length of at least 10 carbon atoms, which was evidence by an increase in zeta potential of TiO_2 particles. Such behavior would suggest cooperative adsorption. Still, the author conducted experiments at a constant alcohol/lignosulfonate ratio, which makes delineating individual contributions difficult, as the lignosulfonate concentration would increase at increasing alcohol levels. Low molecular weight alcohols can indeed increase counterion condensation on lignosulfonates [87], which could facilitate a higher surface coverage. Simon et al. further studied the interfacial tension of lignosulfonate solutions in presence of asphaltenes [82]. The authors concluded that interfacial adsorption was competitive, which is a potential detriment for emulsion stability. At last, Askvik et al. concluded that lignosulfonates and non-ionic surfactants did not associate, due to the small contribution of hydrophobic interaction [38]. Current literate hence disagrees, whether blending lignosulfonates with non-ionic surfactants has a positive effect. Yet, some cases may indeed benefit from such mixtures.

Anionic surfactants can interact with lignosulfonates by electrostatic repulsion, which would suggest competitive adsorption. Still, combining lignosulfonates with other anionic surfactants can be beneficial, as both species increase the ionic strength. As has been discussed previously, increasing salinity will also facilitate more interfacial adsorption [66]. This perception is supported by a study on blending lignosulfonates and sodium dodecyl sulfate (SDS) [75], which showed that the presence of lignosulfonates decreased the critical micelle concentration (CMC) of SDS. A beneficial effect was also found in the context of enhanced oil recovery (EOR) [99–101]. By blending petroleum sulfonates, sodium chloride, 2-propanol, and lignosulfonates, interfacial tension values as low as 1×10^{-3}mN/m were obtained [100]. Another implementation would be the use of lignosulfonates as sacrificial adsorbents [16, 102]. Here, the rock formations are initially saturated with lignosulfonates, after which a second flood with a different surfactant mixture would be injected. While this application does not directly relate to emulsion stabilization, it shows that adsorption was also competitive.

Cationic surfactants interact with lignosulfonates electrostatically, which can yield the formation of lignosulfonate-cationic surfactant complexes [38]. Improved solubility of such complexes in oil media has been reported [103]. Still, a challenge with such systems is the formation of a water-insoluble complexes. In such cases, the surfactants precipitate and are no longer available for interfacial adsorption, hence yielding mixed effects on emulsion stabilization [38].

It has long been established that lignosulfonates can associate with cationic polyelectrolytes, forming insoluble complexes, colloids, and macroscopic precipitates

[104]. Association of lignosulfonates with chitosan reportedly forms such complexes at a sulfonate/amine ratio close to 1.0, suggesting that all sulfonate groups are accessible for interactions [105]. No complex formation was reported at pH 8, which entails that it was indeed electrostatic interactions governing the association of these two compounds. Another interesting application is the formation of multilayers via layer-by-layer association of lignosulfonates and cationic polyelectrolytes [106, 107]. This self-assembly was reportedly governed by electrostatic interactions, hydrogen bonding, and cation-π interactions.

It can be concluded that strong interactions exist between lignosulfonates and cationic surfactants or polymers. Still, these interactions frequently yield precipitates, which would shift the emulsion stabilization mechanism from interfacial adsorption to that of a Pickering emulsion. Beneficial interactions can occur due to mixing lignosulfonates with anionic or non-ionic surfactants, but these may depend on the actual system and the application mode.

4. Emulsion stabilization mechanisms

4.1 Basic concepts

Emulsion stabilization entails the prevention or delay of the coalescence event. Coalescence is the fusing of two or more droplets to form one larger droplet. The limiting case for coalescence is complete phase separation, i.e., the oil and the water being separated into two distinct phases. Coalescence involves film drainage, during which the continuous phase is displaced between the coalescing droplets. Flocculation usually precedes coalescence, in which the dispersed droplets collect to form larger aggregates. Creaming or sedimentation occurs, if the flocculated droplets accumulate at the top (creaming) or the bottom (sedimentation) of the continuous phase. Centrifugation of emulsions stabilized with lignosulfonates can yield the formation of a dense packed layer (DPL) of droplets, which exhibit thixotropy and viscoelastic behavior [13].

Fundamentally speaking, the use of a stabilizer (surfactant and/or polymer) introduces an energy barrier between the droplets [76]. The lowest state of energy would be a system, which is completely phase separated. Yet, the transition from emulsion to complete phase separation may become noncontinuous in presence of a stabilizer. Emulsions stabilized with lignosulfonates are hence only kinetically stable. This entails that with time, the emulsions are expected to return to original state, i.e., a fully coalesced and phase separated system. Still, this transition is hindered to the extent that emulsion can remain in their emulsified state over a period of months or even years.

The interfacial tension is an important parameter, as low interfacial tension reduces the energy required for emulsification. In addition, this parameter can be used to study interfacial phenomena and interactions in the aqueous phase. A known stabilizer tends to be more effective at parameters, which yield a higher reduction of interfacial tension. Still, the reduction of interfacial tension in general is no guarantee for forming stable emulsions. There are examples of surfactants, which substantially decrease the interfacial tension, but do not produce stable emulsions or even destabilize existing systems. The latter are referred to as demulsifiers or emulsion breakers. The important point is that interfacial tension measurements can yield complementary information, but it should not be taken as a sole measure to probe emulsion

DOI: http://dx.doi.org/10.5772/intechopen.107336

stability. Interfacial adsorption is a prerequisite for an efficient stabilizer; however, emulsion stabilization involves several other mechanisms, which will be discussed in detail further on.

4.2 Stearic hindrance

Stearic hindrance or stearic repulsion relies on the presence of the surfactant or polymer at the interface. By imposing spatial obstacles, the oil within the droplets is prevented from coalescing. Stearic repulsion is an important mechanism for non-ionic surfactants and polymers, as these lack the contribution from electrostatic repulsion.

Stearic repulsion is aided by a positive osmotic free energy of interaction, which states that the affinity of the adsorbed species to the continuous phase (water) is greater than to each other [76]. As such, complete film drainage can be prevented, as the surfactant or polymer favors the retention of water between the oil droplets.

A second effect is of entropic nature, also referred to as volume restriction or elastic interaction [76]. A significant overlap of the polymer chains can be favored by hydrophobic or van der Waals interactions. Separating these chains would require energy, which can act as a barrier to prevent coalescence. This phenomenon will be discussed in further detail during Section 4.5.

4.3 Marangoni effect

The Marangoni effect, also referred to as Marangoni-Gibbs effect, is a common effect that facilitates emulsion stabilization with simple surfactants. It is related to the mass transfer along the interface between two fluids due to a concentration gradient. In analogy to Fick's law of diffusion, mass transfer of surfactant molecules is directed towards areas with lower concentration. As the oil droplets approach, the surfactant layer between each droplet and the continuous phase can be displaced. This will in turn yield a concentration gradient. Surfactant molecules are hence drawn back into the contact area between the oil droplets, yielding a stabilizing effect. An illustration of the Marangoni effect is given in **Figure 14**.

Attribution of the Marangoni effect to emulsion stabilization with lignosulfonates is given only implicitly, as the contribution of individual effects is often difficult to delineate. Still, the described effect is likely of importance, as lignosulfonates are subject to interfacial diffusion in the same manner as other surfactants.

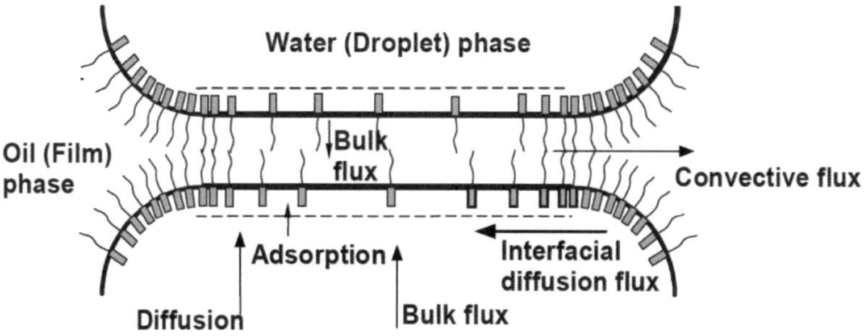

Figure 14.
Illustration of the Marangoni effect during film drainage. Image taken from [108].

4.4 Electrostatic repulsion

In aqueous solution, lignosulfonates can attain a negative charge due to the dissociation of anionic groups. When adsorbed at the interface, these groups will contribute to an overall negative charge. Coulomb forces are then acting between the interfaces of different oil droplets, yielding electrostatic repulsion. This mechanism has been described for stabilization of both particles and emulsions with lignosulfonates [38, 46, 109]. Related phenomena are frequently studied by measuring the electrophoretic mobility of the dispersed particles or droplets, i.e., the zeta potential [38, 70].

Electrostatic repulsion is highly affected by the composition of the aqueous phase. Increasing electrolyte concentrations or the presence of less polar solvents will affect counterion condensation [87]. Charge screening and a lower degree of dissociation can then lessen the repulsion between anionic groups [88]. This could explain the destabilizing effect, which the presence of low molecular weight alcohols can have [81]. Still, increasing the concentration of simple electrolytes tends to improve stability, if the lignosulfonate is not precipitated from solution [66]. This effect is likely attributed to the fact that increasing salinity can enhance other effects, such as increased interfacial adsorption. The pH furthermore affects electrostatic repulsion. It has been reported that higher pH induces larger changes of the zeta potential, as the lignosulfonate macromolecules exhibit a higher degree of ionization [70, 110].

4.5 Viscoelastic interface films

It has long been established that droplet coalescence is affected by the stability of interfacial layers [111]. From a purely mechanical point of view, an emulsion would naturally be more stable, if the dispersed droplets were coated by a rigid interface layer. The term rigid hereby refers to the inelastic behavior, as can also be observed for lignosulfonates [66]. These systems are, however, not entirely rigid. Deformation may occur to some extent, bearing both elastic and viscose contributions, hence the terminology viscoelastic interface films.

Upon deformation, a purely elastic film will store the exerted work as potential energy. If the force is no longer acting, the system will return to the originate state or even oscillate, if the viscose contribution is small or negligible. Viscose forces can be viewed in analogy to friction, since energy is dissipated upon deformation, which cannot be retrieved during relaxation. Based on this, the complex modulus E^* can be modeled as in Eq. (6), i.e., as the sum of the elastic contribution E' and the viscose contribution E''. These contributions are also referred to as the apparent elastic dilatational modulus and the apparent viscous dilatational modulus, respectively, as Eq. (6) is based on the framework of dilatational interfacial rheology. During this measurement technique, a droplet is suspended in a continuous phase in presence of surfactants (see **Figure 9**), while undergoing volume expansion and contraction at the frequency ω. This allows the complex modulus E^* to be determined in terms of the change of interfacial tension $d\sigma(t)$ divided by the change in interface area $d\ln A(t)$.

$$E^*(\omega) = E'(\omega) + iE''(\omega) = \frac{d\sigma(t)}{d\ln A(t)} \qquad (6)$$

Pendant drop tensiometry has the advantage of yielding information on both interfacial tension and interfacial rheology. The technique ceases to function, however, if droplet contraction and expansion is not entirely viscoelastic, as the Young Laplace equation is no longer valid. Other techniques to measure surface rheology include the oscillating barrier method, capillary waves, Langmuir trough, and surface rotational shear rheometry [112]. The latter two have been successfully applied for studying the interfacial behavior of lignosulfonates.

Results from interfacial shear rheology of lignosulfonates are depicted in **Figure 15**. Frequency sweeps are of interest, as these can be used to characterize the interfacial properties. As can be seen, the elastic modulus *E'* and the viscose modulus *E"* are intersecting or approaching each other in cases, where only monovalent cations are present. The interfacial response is hence entirely viscoelastic. In presence of multivalent cations, i.e., Ca^{2+} and Al^{3+}, the elastic and viscose modulus are approximately parallel to each other. This behavior is characteristic for that of gelled interfaces. It appears that multivalent cations can function as connectors between the lignosulfonate molecules, imposing a stronger cohesion to the interface film. This gelling is a reasonable explanation for the observed rigidity in **Figure 9**.

Increasing the lignosulfonate concentration has shown to decrease the interfacial tension (see **Figure 8**). While the same could be expected for the interfacial modulus, experimental evidence shows that this is not the case. As depicted in **Figure 16**, the interfacial modulus will increase to a maximum, after which a decrease is noted. It appears that electrolytic effects govern this behavior, as the same maximum was observed for adding simple electrolytes. The maximum has indeed been correlated with a maximum in emulsion stability and the onset of lignosulfonate precipitation [66]. The observed phenomena are hence in line with two long established principles: Emulsions can be stabilized by the formation of viscoelastic interface layers and the emulsion stability tends to be best, if the stabilizing agent is on the verge of precipitation [90, 111]. If the solubility limit is exceeded, precipitation yields the formation of

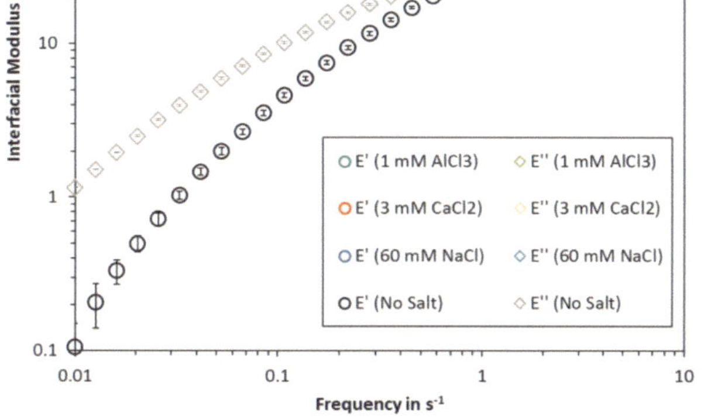

Figure 15.
Frequency sweep of 1.8 g/l sodium lignosulfonate and in presence of different electrolytes at 0.2% strain. The oil phase is made of xylene isomer blend. Figure taken from [66].

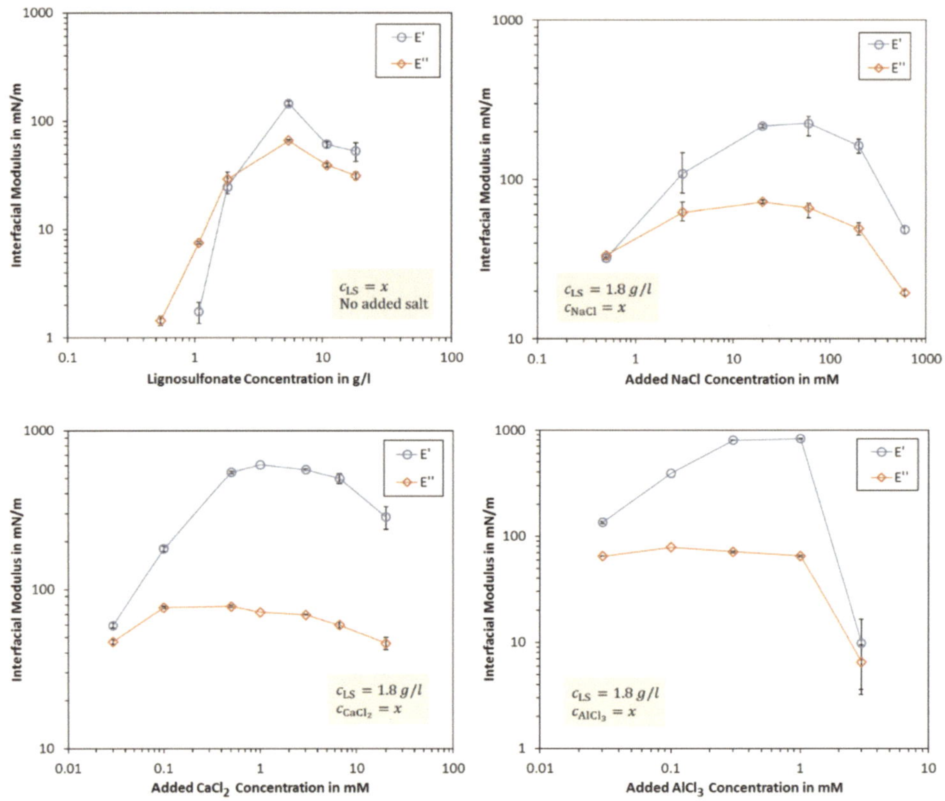

Figure 16.
Interfacial moduli in dependence of lignosulfonate or added salt concentration. Data was obtained using interfacial shear rheology and xylene isomer blend as the oil phase. Figures taken from [66].

particles, hence shifting the stabilization mechanism to that of a Pickering emulsion. This type of mechanism will be discussed in the next sub-chapter.

4.6 Particle stabilization

Particle stabilization of emulsions involves the presence of a third (solid) phase. The solid phase is usually present as a colloidal dispersion, exhibiting particle sizes of approximately 1 nm to 1 μm. These particles can adsorb at the oil–water interface, forming bilayers and bridging monolayers [113]. The stabilizing mechanism is based on coherent particle layers around the dispersed liquid, preventing coalescence by acting as a steric (mechanical) barrier. A particle stabilized emulsion is also referred to as Pickering emulsion. Interest has currently shifted from using inorganic particles to adopting bio-based systems, which includes the use of cellulose, chitin, starch, proteins, and lignin [114].

Particle stabilization is usually not the primary mechanism, with which lignosulfonates stabilize emulsions. In aqueous solution at low salinity, lignosulfonates are usually well-dissolved and hence act by molecular adsorption at the interface. Certain conditions, however, can induce lignosulfonate precipitation, which will shift the stabilization mechanism to that of a Pickering emulsion [66]. In addition, lignosulfonate aggregation can occur at concentrations as low as 0.05 g/l [55]. Aggregation is reportedly facilitated by π–π stacking, hydrogen bonding, and hydrophobic interactions [93, 115, 116]. At the right conditions, aggregate dimensions can be in the range of colloidal dispersions. Based on the diffusion coefficient of lignosulfonates obtained from dilatational interfacial rheology, it was concluded that lignosulfonates undergo interfacial adsorption in the aggregated state [66]. Interfacially adsorbed aggregates could furthermore act by particle stabilization. It is important to note that lignosulfonate aggregation occurs gradually over a broad concentration range. The contribution of aggregates to the overall emulsion stability would hence build up in increments. This contrasts with precipitation, where an immediate shift to particle stabilization is observed at the precipitation onset.

5. Summary and conclusion

This chapter detailed the fundamentals of emulsion stabilization with lignosulfonates. First, basic concepts, the industrial production, as well as chemical make-up and structure of lignosulfonates were discussed. Second, the effect of lignosulfonates on surface and interfacial tension was described. Third, the fundamental mechanisms behind emulsion stabilization with lignosulfonates were explained.

Lignosulfonates are bio-based specialty chemicals and function as surfactants due to their amphiphilic property. Lignosulfonates are readily water-soluble as they comprise ionizable moieties, i.e., sulfonate and carboxylic acid groups at neutral pH. They can hence be utilized in a variety of technical applications, which includes the stabilization of oil-in-water emulsions. Surface and interface adsorption of lignosulfonates is evidenced by a reduction in surface and interfacial tension. Parameters that can enhance the effect on interfacial tension include increased salinity, a reduction in pH, and the presence of co-solvents or co-surfactants. A high reduction of interfacial tension is generally beneficial, as this suggests enhanced interfacial adsorption and reduces the energy required during emulsification. While interfacial adsorption is a prerequisite for an efficient emulsion stabilizer, the stabilization mechanism is related to other phenomena as well. For lignosulfonates, these mechanisms include stearic hindrance, electrostatic repulsion, the Marangoni-Gibbs effect, the formation of viscoelastic interface layers, and particle stabilization. The stabilization mechanisms are furthermore affected by the composition of the lignosulfonates as well as the aqueous and oil phase. Parameters that increase the interfacial activity and yield the formation of more cohesive interface layers are beneficial, as both effects tend to promote emulsion stability.

There has been an increased interest in biopolymers recently, as these are inherently more sustainable than petroleum-derived polymers. Lignosulfonates are no exception to that trend, which is also mirrored by the large variety of lignosulfonate products on the market today. Emulsion stabilization is one of the application areas, which still has the potential for growth. This chapter was hence dedicated to the fundamentals of emulsion stabilization with lignosulfonates, in the hope that this may aid their utilization in new areas and products.

Conflict of interest

The author declares no conflict of interest.

Author details

Jost Ruwoldt
RISE PFI AS, Trondheim, Norway

*Address all correspondence to: jostru.chemeng@gmail.com

References

[1] Sixta H. Handbook of Pulp. Weinheim, Germany: Wiley-vch; 2006

[2] Ruwoldt J. A critical review of the physicochemical properties of lignosulfonates: Chemical structure and behavior in aqueous solution, at surfaces and interfaces. Surfaces. 2020;**3**(4): 622-648

[3] Vishtal AG, Kraslawski A. Challenges in industrial applications of technical lignins. BioResources. 2011;**6**(3): 3547-3568

[4] Scamehorn, J.F., An Overview of Phenomena Involving Surfactant Mixtures. United States: 1986.

[5] Aro T, Fatehi P. Production and application of lignosulfonates and sulfonated lignin. ChemSusChem. 2017; **10**(9):1861-1877

[6] Macfarlane AL, Mai M, Kadla JF. In: Waldron K, editor. 20—Bio-Based Chemicals from Biorefining: Lignin Conversion and Utilisation, Advances in Biorefineries. Amsterdam, Boston, Cambridge, Heidelberg, London, New York, Oxford, Paris, San Diego, San Francisco, Singapore, Sydney, Tokyo: Woodhead Publishing; 2014. pp. 659-692

[7] Stern T, Schwarzbauer P. Wood-based lignosulfonate versus synthetic polycarboxylate in concrete admixture systems: The perspective of a traditional pulping by-product competing with an oil-based substitute in a business-to-business market in Central Europe. Forest Products Journal. 2008;**58**:81-86

[8] Lauten RA, Myrvold BO, Gundersen SA. New developments in the commercial utilization of lignosulfonates. Surfactants from Renewable Resources. Chichester (UK): John Wiley & Sons, Ltd; 2010. pp. 269-283

[9] Xu C, Ferdosian F. Utilization of lignosulfonate as dispersants or surfactants. In: Xu C, Ferdosian F, editors. Conversion of Lignin into Bio-Based Chemicals and Materials. Berlin Heidelberg: Springer; 2017. pp. 81-90

[10] Garguiak JD, Lebo SE. Commercial use of lignin-based materials. In: Lignin: Historical, Biological, and Materials Perspectives. Washington DC: American Chemical Society; 1999. pp. 304-320

[11] Miretzky P, Cirelli AF. Cr(VI) and Cr(III) removal from aqueous solution by raw and modified lignocellulosic materials: A review. Journal of Hazardous Materials. 2010;**180**(1):1-19

[12] Alazigha DP et al. Mechanisms of stabilization of expansive soil with lignosulfonate admixture. Transportation Geotechnics. 2018;**14**:81-92

[13] Ruwoldt J, Planque J, Øye G. Lignosulfonate salt tolerance and the effect on emulsion stability. ACS Omega. 2020;**5**(25):15007-15015

[14] Ouyang X et al. Corrosion and scale inhibition properties of sodium lignosulfonate and its potential application in recirculating cooling water system. Industrial & Engineering Chemistry Research. 2006;**45**(16):5716-5721

[15] Blanck G, Cuisinier O, Masrouri F. Soil treatment with organic non-traditional additives for the improvement of earthworks. Acta Geotechnica. 2014;**9**(6):1111-1122

[16] Tsau J-S et al. Use of sacrificial agents in CO_2 foam flooding application. In: SPE Annual Technical Conference

and Exhibition. Houston, TX: Society of Petroleum Engineers; 1999. p. 9

[17] Chiwetelu C. Improving the oil recovery efficacy of lignosulfonate solutions. Journal of Canadian Petroleum Technology. 1980;**19**(03):10

[18] Vebi M et al. Lignosulfonate-based polyurethane materials via cyclic carbonates: Preparation and characterization. Holzforschung. 2020; **74**(2):203-211

[19] Hirose S. Novel epoxy resins with unsaturated ester chains derived from sodium lignosulfonate. Macromolecular Symposia. 2015;**353**(1):31-38

[20] Hu J-P, Guo M-H. Influence of ammonium lignosulfonate on the mechanical and dimensional properties of wood fiber biocomposites reinforced with polylactic acid. Industrial Crops and Products. 2015;**78**:48-57

[21] Boerjan W, Ralph J, Baucher M. Lignin biosynthesis. Annual Review of Plant Biology. 2003;**54**(1):519-546

[22] Myrvold BO. A new model for the structure of lignosulphonates: Part 1. Behaviour in dilute solutions. Industrial Crops and Products. 2008;**27**(2):214-219

[23] de Jong E, Higson A, Walsh P, Wellisch M. Bio-based chemicals value added products from biorefineries. IEA Bioenergy, Task42 Biorefinery, 2012:34

[24] Naik SN et al. Production of first and second generation biofuels: A comprehensive review. Renewable and Sustainable Energy Reviews. 2010;**14**(2): 578-597

[25] Laurichesse S, Avérous L. Chemical modification of lignins: Towards biobased polymers. Progress in Polymer Science. 2014;**39**(7):1266-1290

[26] Johansson A, Aaltonen O, Ylinen P. Organosolv pulping—Methods and pulp properties. Biomass. 1987;**13**(1):45-65

[27] Brosse N, Hussin MH, Rahim AA. Organosolv processes. Biorefineries. 2017;**166**:153-176

[28] Ruwoldt J, Tanase Opedal M. Green materials from added-lignin thermoformed pulps. Industrial Crops and Products. 2022;**185**:115102

[29] Yamamoto M. St1 Cellunolix® process–Lignocellulosic bioethanol production and value chain upgrading. In: Bio4Fuels Days. Oslo (Norway): Oral Presentation; 2018

[30] Pitk P. Wood fractionation scale-up for sustainable raw materials. In: Lignocity Webinars: Oral Presentation; 2021

[31] Shevchenko SM, Beatson RP, Saddler JN. The nature of lignin from steam explosion/enzymatic hydrolysis of softwood. In: Twentieth Symposium on Biotechnology for Fuels and Chemicals. Amsterdam, Boston, Cambridge, Heidelberg, London, New York, Oxford, Paris, San Diego, San Francisco, Singapore, Sydney, Tokyo: Springer; 1999

[32] Sarker TR et al. Hydrothermal pretreatment technologies for lignocellulosic biomass: A review of steam explosion and subcritical water hydrolysis. Chemosphere. 2021;**284**:131372

[33] Pu Y, Jiang N, Ragauskas AJ. Ionic liquid as a green solvent for lignin. Journal of Wood Chemistry and Technology. 2007;**27**(1):23-33

[34] Szalaty TJ, Klapiszewski Ł, Jesionowski T. Recent developments in modification of lignin using ionic liquids for the fabrication of advanced

materials—A review. Journal of Molecular Liquids. 2020;**301**:112417

[35] Gil-Chávez J et al. Optimization of the spray-drying process for developing aquasolv lignin particles using response surface methodology. Advanced Powder Technology. 2020;**31**(6):2348-2356

[36] Daintith J. Surfactant. New York: Oxford University Press; 2008

[37] Paria S, Khilar KC. A review on experimental studies of surfactant adsorption at the hydrophilic solid–water interface. Advances in Colloid and Interface Science. 2004;**110**(3):75-95

[38] Askvik KM et al. Complexation between lignosulfonates and cationic surfactants and its influence on emulsion and foam stability. Colloids and Surfaces A: Physicochemical and Engineering Aspects. 1999;**159**(1):89-101

[39] Ingruber OV, Kocurek MJ, Wong A. Sulfite Science and Technology, In: Pulp and Paper Manufacture Series, 4. Unites States: TAPPI Press; 1997

[40] Nimz HH et al. Carbon-13 NMR spectra of lignins, 8. Structural differences between lignins of hardwoods, softwoods, grasses and compression wood. Holzforschung—International Journal of the Biology, Chemistry, Physics and Technology of Wood. 1981;**35**:16

[41] Strassberger Z et al. Lignin solubilisation and gentle fractionation in liquid ammonia. Green Chemistry. 2015; **17**(1):325-334

[42] Fredheim GE, Christensen BE, Braaten SM. Comparison of molecular weight and molecular weight distributions of softwood and hardwood lignosulfonates. Journal of Wood Chemistry and Technology. 2003;**23**(2): 197-215

[43] Windeisen E, Wegener G. In: Matyjaszewski K, Möller M, editors. 10.15—Lignin as Building Unit for Polymers, in Polymer Science: A Comprehensive Reference. Amsterdam: Elsevier; 2012. pp. 255-265

[44] Bhattacharya P et al. Studies on ultrafiltration of spent sulfite liquor using various membranes for the recovery of lignosulphonates. Desalination. 2005;**174**(3):287-297

[45] Howard GC. Utilization of sulfite, liquor. Industrial & Engineering Chemistry. 1934;**26**(6):614-617

[46] Subramanian S, Øye G. Aqueous carbon black dispersions stabilized by sodium lignosulfonates. Colloid and Polymer Science. 2021;**299**(7): 1223-1236

[47] El Mansouri N-E, Salvadó J. Analytical methods for determining functional groups in various technical lignins. Industrial Crops and Products. 2007;**26**(2):116-124

[48] Mansouri N-EE, Salvadó J. Structural characterization of technical lignins for the production of adhesives: Application to lignosulfonate, Kraft, soda-anthraquinone, organosolv and ethanol process lignins. Industrial Crops and Products. 2006;**24**(1):8-16

[49] Lugovitskaya T, Bolatbaev K, Naboichenko S. Study of surface phenomena at phase boundaries in the presence of lignosulfonates. Russian Journal of Applied Chemistry. 2012; **85**(8):1192-1196

[50] Brovko O et al. Composite aerogel materials based on lignosulfonates and silica: Synthesis, structure, properties. Materials Chemistry and Physics. 2021; **269**:124768

[51] Kun D, Pukánszky B. Polymer/lignin blends: Interactions, properties, applications. European Polymer Journal. 2017;**93**:618-641

[52] Fiorani G et al. Advancements and complexities in the conversion of lignocellulose into chemicals and materials. Frontiers in Chemistry. 2020; **8**:797

[53] Brauns FE, Brauns DA. The Chemistry of Lignin Supplement Volume Covering the Literature for the Years 1949-1958. New York and London: Academic Press; 1960

[54] Deng Y et al. Adsorption and desorption behaviors of lignosulfonate during the self-assembly of multilayers. BioResources. 2010;**5**(2):1178-1196

[55] Yan M et al. Influence of pH on the behavior of lignosulfonate macromolecules in aqueous solution. Colloids and Surfaces A: Physicochemical and Engineering Aspects. 2010;**371**(1):50-58

[56] Belyi V et al. Study of acid-base properties of lignin using the method of pK-spectroscopy. Butlerov Communications. 2013;**35**(7):102-108

[57] Laughlin RG. HLB, from a thermodynamic perspective. Journal of the Society of Cosmetic Chemists. 1981; **32**:371-392

[58] Ensing B et al. On the origin of the extremely different solubilities of polyethers in water. Nature Communications. 2019;**10**(1):2893

[59] Holmberg K. Natural surfactants. Current Opinion in Colloid & Interface Science. 2001;**6**(2):148-159

[60] Eraghi Kazzaz A, Hosseinpour Feizi Z, Fatehi P. Grafting strategies for hydroxy groups of lignin for producing materials. Green Chemistry. 2019; **21**(21):5714-5752

[61] Lebo SE, et al. Recent advances in the characterization of lignosulfonates, in Characterization of Lignocellulosic Materials, T.Q. Hu, editor. Oxford (UK), Ames (USA), Carlton (Australia): Blackwell Publishing Ltd.: 2008

[62] Ekeberg D et al. Characterisation of lignosulphonates and Kraft lignin by hydrophobic interaction chromatography. Analytica Chimica Acta. 2006;**565**(1):121-128

[63] Winowiski T et al. Characterization of sulfonated lignin dispersants by hydrophobic interactive chromatography. Journal of ASTM International. 2005;**2**(9):1-6

[64] Musl O et al. Hydrophobic interaction chromatography (HIC) in 2D-LC characterization of lignosulfonates. ChemSusChem. 2021;**9** (49):16786-16795

[65] Kontturi A-K. Diffusion coefficients and effective charge numbers of lignosulphonate. Influence of temperature. Journal of the Chemical Society, Faraday Transactions 1: Physical Chemistry in Condensed Phases. 1988; **84**(11):4043-4047

[66] Ruwoldt J, Simon S, Øye G. Viscoelastic properties of interfacial lignosulfonate films and the effect of added electrolytes. Colloids and Surfaces A: Physicochemical and Engineering Aspects. 2020;**606**:125478

[67] Fredheim GE, Braaten SM, Christensen BE. Molecular weight determination of lignosulfonates by size-exclusion chromatography and multi-angle laser light scattering. Journal of Chromatography A. 2002;**942**(1):191-199

[68] Ge Y et al. Influence of molecular mass of lignosulfonates on the resulting surface charges of solid particles. International Journal of Biological Macromolecules. 2013;**52**:300-304

[69] Tolbert A et al. Characterization and analysis of the molecular weight of lignin for biorefining studies. Biofuels, Bioproducts and Biorefining. 2014;**8**(6): 836-856

[70] Ratinac KR, Standard OC, Bryant PJ. Lignosulfonate adsorption on and stabilization of lead zirconate titanate in aqueous suspension. Journal of Colloid and Interface Science. 2004;**273**(2): 442-454

[71] Zulfikar MA, Wahyuningrum D, Lestari S. Adsorption of lignosulfonate compound from aqueous solution onto chitosan-silica beads. Separation Science and Technology. 2013;**48**(9):1391-1401

[72] Bai B, Grigg RB. Kinetics and equilibria of calcium lignosulfonate adsorption and desorption onto limestone. In: SPE International Symposium on Oilfield Chemistry. The Woodlands, TX: Society of Petroleum Engineers; 2005. p. 11

[73] Pang Y-X et al. Influence of oxidation, hydroxymethylation and sulfomethylation on the physicochemical properties of calcium lignosulfonate. Colloids and Surfaces A: Physicochemical and Engineering Aspects. 2008;**312**(2):154-159

[74] Langmuir I. The constitution and fundamental properties of solids and liquids. Part I. Solids. Journal of the American Chemical Society. 1916;**38**(11): 2221-2295

[75] Rana D, Neale G, Hornof V. Surface tension of mixed surfactant systems: Lignosulfonate and sodium dodecyl sulfate. Colloid and Polymer Science. 2002;**280**(8):775-778

[76] Tadros TF. Emulsion formation, stability, and rheology. Emulsion Formation and Stability. 2013;**1**:1-75

[77] Bai B, Wu Y, Grigg RB. Adsorption and desorption kinetics and equilibrium of calcium lignosulfonate on dolomite porous media. The Journal of Physical Chemistry C. 2009;**113**(31):13772-13779

[78] Zulfikar MA, Setiyanto H, Djajanti SD. Effect of temperature and kinetic modelling of lignosulfonate adsorption onto powdered eggshell in batch systems. Songklanakarin Journal of Science and Technology. 2013;**35**:1-31

[79] Klapiszewski Ł et al. Silica/ lignosulfonate hybrid materials: Preparation and characterization. Central European Journal of Chemistry. 2014;**12**(6):719-735

[80] Li H-q, et al. Kinetic and equilibrium studies on the adsorption of calcium lignosulfonate from aqueous solution by coal fly ash. Chemical Engineering Journal. 2012;**200-202**:275-282

[81] Ruwoldt J, Øye G. Effect of low-molecular-weight alcohols on emulsion stabilization with lignosulfonates. ACS Omega. 2020;**5**(46):30168-30175

[82] Simon S et al. Lignosulfonates in crude oil processing: Interactions with asphaltenes at the oil/water Interface and screening of potential applications. ACS Omega. 2020;**5**(46):30189-30200

[83] Verruto VJ, Le RK, Kilpatrick PK. Adsorption and molecular rearrangement of amphoteric species at oil–water interfaces. The Journal of Physical Chemistry B. 2009;**113**(42): 13788-13799

[84] Freer EM, Radke CJ. Relaxation of asphaltenes AT the toluene/water interface: Diffusion exchange and surface rearrangement. The Journal of Adhesion. 2004;**80**(6):481-496

[85] Keleşoğlu S et al. Rheological properties of highly concentrated dense packed layer emulsions (w/o) stabilized by asphaltene. Journal of Petroleum Science and Engineering. 2015;**126**:1-10

[86] Qiu X et al. Effect of straight-chain alcohols on the physicochemical properties of calcium lignosulfonate. Journal of Colloid and Interface Science. 2009;**338**(1):151-155

[87] Vainio U et al. Distribution of counterions around lignosulfonate macromolecules in different polar solvent mixtures. Langmuir. 2012;**28**(5): 2465-2475

[88] Myrvold BO. The polyelectrolyte behavior of randomly branched lignosulfonates. Tappi Journal. 2007; **6**(11):10-14

[89] Myrvold BO. Salting-out and salting-in experiments with lignosulfonates (LSs). Holzforschung. 2013;**67**(5):549-557

[90] Nielsen LE, Wall R, Adams G. Coalescence of liquid drops at oil-water interfaces. Journal of Colloid Science. 1958;**13**(5):441-458

[91] Yan M, Yang D. Adsorption mechanism of lignosulfonate at the air/liquid interface. Journal of the Brazilian Chemical Society. 2015;**26**(3):555-561

[92] Park S, Lee ES, Sulaiman WRW. Adsorption behaviors of surfactants for chemical flooding in enhanced oil recovery. Journal of Industrial and Engineering Chemistry. 2015;**21**: 1239-1245

[93] Li H et al. Effect of temperature on polyelectrolyte expansion of lignosulfonate. BioResources. 2015; **10**(1):575-587

[94] Qian Y et al. Aggregation of sodium lignosulfonate above a critical temperature. Holzforschung. 2014; **68**(6):641-647

[95] Österberg M et al. Spherical lignin particles: A review on their sustainability and applications. Green Chemistry. 2020;**22**(9):2712-2733

[96] Americas ICI. The HLB system: A time-saving guide to emulsifier selection. Wilmington (USA): ICI Americas, Incorporated; 1984

[97] Setiati R et al. Challenge sodium lignosulfonate surfactants synthesized from bagasse as an injection fluid based on hydrophil liphophilic balance. In: IOP Conference Series: Materials Science and Engineering. Bristol (UK): IOP Publishing; 2018

[98] Graciaa A et al. Improving solubilization in microemulsions with additives. 2. Long chain alcohols as lipophilic linkers. Langmuir. 1993;**9**(12): 3371-3374

[99] Manasrah K, Neale G, Hornof V. Properties of mixed surfactant solutions containing petroleum sulfonates and lignosulfonates. Cellulose Chemistry and Technology. 1985**19**(3):291-299

[100] Hornof V et al. Lignosulfonate-based mixed surfactants for low interfacial tension. Cellulose Chemistry and Technology. 1984;**18**(3):297-303

[101] Hong SA, Bae JH, Lewis GR. An evaluation of lignosulfonate as a sacrificial adsorbate in surfactant flooding. SPE Reservoir Engineering. 1987;**2**(01):17-27

[102] Hong SA, Bae JH. Field experiment of lignosulfonate preflushing for surfactant adsorption reduction. SPE Reservoir Engineering. 1990;**5**(04): 467-474

[103] Askvik KM et al. Properties of the lignosulfonate–surfactant complex phase. Colloids and Surfaces A: Physicochemical and Engineering Aspects. 2001;**182**(1):175-189

[104] Ström G, Stenius P. Formation of complexes, colloids and precipitates in aqueous mixtures of lignin sulphonate and some cationic polymers. Colloids and Surfaces. 1981;**2**(4):357-371

[105] Fredheim GE, Christensen BE. Polyelectrolyte complexes: Interactions between lignosulfonate and chitosan. Biomacromolecules. 2003;**4**(2):232-239

[106] Ouyang X et al. Adsorption characteristics of lignosulfonates in salt-free and salt-added aqueous solutions. Biomacromolecules. 2011;**12**(9): 3313-3320

[107] Luo H et al. Structure and properties of layer-by-layer self-assembled chitosan/lignosulfonate multilayer film. Materials Science and Engineering: C. 2012;**32**(7):2001-2006

[108] Spiecker PM. The Impact of Asphaltene Chemistry and Solvation on Emulsion and Interfacial Film Formation. Raleigh (USA): North Carolina State University; 2001

[109] Nanthakumar B, Arinaitwe E, Pawlik M. Adsorption of sodium lignosulfonates on hematite. Adsorption. 2010;**16**(4):447-455

[110] Megiatto JD, Cerrutti BM, Frollini E. Sodium lignosulfonate as a renewable stabilizing agent for aqueous alumina suspensions. International Journal of Biological Macromolecules. 2016;**82**:927-932

[111] Biswas B, Haydon D. The coalescence of droplets stabilised by viscoelastic adsorbed films. Kolloid-Zeitschrift und Zeitschrift für Polymere. 1962;**185**(1):31-38

[112] Derkach SR, Krägel J, Miller R. Methods of measuring rheological properties of interfacial layers (Experimental methods of 2D rheology). Colloid Journal. 2009;**71**(1):1-17

[113] Horozov TS, Binks BP. Particle-stabilized emulsions: A bilayer or a bridging monolayer? Angewandte Chemie International Edition. 2006; **45**(5):773-776

[114] Lam S, Velikov KP, Velev OD. Pickering stabilization of foams and emulsions with particles of biological origin. Current Opinion in Colloid & Interface Science. 2014;**19**(5):490-500

[115] Deng Y et al. Pi-pi stacking of the aromatic groups in lignosulfonates. BioResources. 2012;**7**(1):1145-1156

[116] Vainio U, Lauten RA, Serimaa R. Small-angle X-ray scattering and rheological characterization of aqueous lignosulfonate solutions. Langmuir. 2008;**24**(15):7735-7743

Chapter 5

Fractionation of Lignin for Valorization

Zhenglun Li

Abstract

Biorefining produces technical lignins that are not readily usable as a precursor for the production of value-added materials and chemicals, thus yielding a technological gap in complete utilization of lignocellulosic biomass. Various processes have been developed and demonstrated for fractionation of technical lignins, with the purpose of increasing the suitability of technical lignins as a precursor for commercial production of fuels and chemicals. Fractionation of lignins reduces the amount of impurities and generates lignin streams with smaller internal variations in chemical and structural properties. Examples of such processes, including membrane filtration, solvent extraction, and acid gradient precipitation, are reviewed in this chapter.

Keywords: lignin, liquid–liquid extraction, membrane separation, biorefining

1. Introduction

Lignin is a loosely defined term that is used to describe different types of plant-based materials depending on the context. Lignin in different plants can be different in monolignol composition and molecular structure. On top of this variability is the susceptibility of lignins to various lignin extraction processes, as well as the various chemical and structural changes of extracted lignins during processing. In this chapter, the stream of lignins derived from a biorefining process (e.g., kraft pulping, cellulosic ethanol production) are referred to as "technical lignins", which are compositionally and structurally different from the native lignins present in plant cell walls.

Depending on the isolation process, technical lignins contain components other than phenolic polymers. Some of these components (e.g., sodium salt, solvents, carbohydrates, extractives) may complicate the downstream processing of value-added products. The most common types of technical lignin are kraft lignin and sulfite lignin, of which 5×10^7 tons are produced annually by the pulp and paper industry [1]. These lignins contain partially depolymerized plant lignins of various molecular weight, along with sodium salts as well as sulfur-containing organic and inorganic components. Lignins derived from cellulosic biorefining (e.g., biochemical conversion of lignocellulose feedstocks via saccharification and fermentation) contain ash components, undigested plant polysaccharides as well as oligomeric and monomeric sugars [2, 3]. These non-lignin components have various influences on lignin valorization. For instance, ash components introduce defects and negatively affects the crystallinity of carbon fiber produced from lignins. Sugars and carbohydrates participate in condensation reactions with lignins under thermal treatment, which complicate the processing of lignins into thermoplastic materials.

Fractionation of technical lignins is a necessary step in the upgrading of lignins to high-value products, which has higher requirements in product purity and/or performance compared to other lower-value uses of technical lignins (e.g., boiler fuel, dust control agent). In the conversion process of technical lignins to lignin fiber and carbon fiber, soften spinning lignin is complicated by the high polydispersity of and branching of technical lignins [4]. As for catalytic hydrogenolysis of technical lignins, the noble metal catalysts lose activity in presence of the sulfur impurities in technical lignins [5]. With fractionation processes, these adverse effects of structural heterogeneity and chemical impurity can be addressed as suitable precursors for valorization is derived from technical lignins.

2. pH-based precipitation

Lignins obtained via kraft pulping processes have been commercially separated from black liquor via pH-based precipitation. As pH decreases, lignin molecules become protonated and have less intermolecular electrostatic repulsion, thus favoring the formation of lignin precipitates. Sequential pH adjustment has been studied as a process for producing kraft lignin fractions with different molecular weights, different chemical structures, and different ash content. Magalhães et al. studied a sequential acidification process and obtained low-ash, fully protonated lignins at pH 3 from kraft lignins derived from both hardwood and softwood feedstocks [6]. Helander and Theliander reported the positive correlation between precipitation pH and molecular weight in the recovered fractions, and the retention of most sulfur content in the low molecular weight lignin fraction [7].

Lignin precipitation is also affected by anionic strength, and the effects of different anions on lignin prepitation follow the same order as the Hofmeister series [8]. Sewring and Theliander compared the difference between chloride and sulfate anions regarding their effect on lignin precipitation, and discovered that sulfates not being an effective anion for salting out kraft lignins [9]. When carbon dioxide is used in lignin precipitation, lignin precipitates can be formed with low ash content (<1 wt%) [10]. This process has been commercialized (e.g., LignoBoost) and shows a potential as a sustainable technology given the availability of renewable CO_2.

3. Solvent extraction

Solvent extraction, e.g., partitioning and extraction with organic solvent, has been investigated both as a method of isolating lignins from biomass (for production of Organosolv lignins and Acetosolv lignins) and as a process for fractionating technical lignins. The structure and composition of extracted lignins is affected by the type of solvent used and extraction process parameters. Saddler et al. reported the lignins extracted via an ethanol organosolv process and obtained lignins with Mw $\approx$ 2100 and a polydispersity (PDI) of 1.8–2.0 [11], which are similar to those of organosolv lignins obtained from Alcell (Mw = 2100–8000, PDI = 3.5–13) and methanol organosolv (Mw = 1200–3800, PDI = 1.6–2.4) processes [12]. With increasing severity of the extractions, the average molecular weight of the extracted lignins decreases and the number of functional groups in the extracted lignins decreases. The increase in functional groups changes the lignins' reactivity during derivatization reactions, thus improving their suitability in upgrading processes via chemical modification.

Extracted fractions of technical lignins are also used in the synthesis of lignin derivatives and copolymers. Generally speaking, organic solvent extracted lignins have lower average molecular weight and lower glass transition temperature compared to the lignins prior to the extraction [13, 14]. Organosolv lignin (e.g., Alcell), which is extracted from hardwood using a mixture of organic solvents, has been demonstrated as a viable precursor for the synthesis of polyols [15] and lignin esters [16, 17]. Pan and Saddler compared the performance of organosolv and Kraft lignins as polyol substitutes in the systhesis of rigid polyurethane foam, and the results showed organosolv lignin being a better option with greater miscibility with commercially-available synthetic polyols [18]. The amount of polyol groups in organosolv lignin can be affected by the type of organic solvent used in the extraction process, thus enabling some degree of tuning in the reactivity of extracted lignins [14, 19].

Solvent fractionation of technical lignins removes inorganic salts and retains the low molecular weight fractions of lignins, thus improving the suitability of lignins as a precursor of melt spinning in carbon fiber production. Baker et al. reported the use of a solvent extraction process in obtaining a lignin fraction that has undetectable level of ash (compared to 2.7 wt% of ash content in the hardwood lignin prior to the extraction), low crosslinking reactivity (thus forming a more stable melt), and easier extrusion [20]. Rials et al. explored the use of organosolv lignins obtained from extraction of switchgrass and yellow poplar using an acidified solvent mixture (i.e. a mixture of methyl isobutyl ketone, ethanol, and water), and discovered that the blend of lignins from the two biomass feedstocks demonstrated better mechanical properties compared to organosolv lignins solely derived from switchgrass [21], due to the plasticizing effect of hardwood lignins. It is worth noting that for fractionated lignins to be a suitable precursor for carbon fiber production, it has to have high purity (<1000 ppm of ash), low volatile content (<5 wt % at 250°C), and low content of non-volatile particulates (<500 ppm) [22]. These requirements can be met with organic solvent extraction processes which aims to dissolve polymeric lignins while partitioning lignins from impurities.

Extraction with low toxicity and renewable solvent mixtures have been identified as a promising approach to fractionating and purifying lignins. Thies et al. reported the use of the mixture of acetic acid and water as a solvent for removing metal salts from kraft lignin under elevated temperature (95°C). Upon mixing the hot acetic acid–water mixture with kraft lignin, two phases are formed: one solvent-rich phase with extracted metal salts and one lignin-rich phase with <100 ppm metals content [23]. Hodge and Thies described a modified version of the aforementioned extraction process, where pressurized carbon dioxide is introduced to the lignin-acid-water mixture to produce a lignin solution in a CO_2-expanded solvent [24]. By controlling the pressure of carbon dioxide, lignin fractions with various average molecular weights can be obtained from kraft lignin. In addition to concentrated acetic acid aqueous solution, deep eutectic organic solvents with low vapor pressures are also reported as candidates for solvent fractionation of lignins, as summarized in a review by Ragauskas and Wan [25]. Hou et al. studied the use of solvents composed of choline chloride and urea (or oxalic acid) in the dissolution of kraft lignins, and obtained lignin solutions with 10–16% w/w concentrations [26]. These eutectic solvents are also used in the extraction of lignins from bamboo [27], corncob [28], and poplar [29] biomass. Due to the high costs with the production and recovery of deep eutectic solvents, this approach to lignin fractionation has not yet been commercialized.

4. Flocculation and membrane separation

The use of flocculants increases the Zeta potential of black liquor and promotes lignin flocculation. Piazza and Garcia studied the application of bovine blood and poly(diallyldimethylammonium chloride) in the recovery of soda lignin derived from wheat straw, and both flocculants are effective (87–92%) in removing lignin from the supernatant [30].

Ultrafiltration is the most commonly used commercial process that recovers lignosulfonates during the sulfite pulping process, and is also used in the recovery of lignins from kraft pulping black liquor. Via ultrafiltration and nanofiltration with membranes of different cut-off, various lignin fractions can be obtained from black liquors. The applications of membrane filtration in lignin recovery has been reviewed by Humpert and Czermak [1]. After partial depolymerization (via pulping and/or other chemical processes), technical lignins can be fractionated via membrane separation and thus the low- and high-molecular-weight fractions can be utilized for different applications. For instance, Hulteburg et al. demonstrated the use of polymeric membrane to recover low molecular weight phenolics from softwood kraft lignins that have been partially depolymerized with aqueous NaOH [31]. The recovered phenolic compounds have a molecular weight of 250–450 g/mol and can be metabolized by some *Pseudomonas* strains [32, 33]. Mendes et al. reported the use of ultrafiltration to recover 83% of the lignins in the lignin-lean aqueous phase derived from the acidification of kraft black liquor [34]. Labidi et al. compared the performance of several ceramic ultrafiltration membranes with different cut-offs (5–15 kDa) in the fractions of soda pulping black liquor from Miscanthus, and discovered that other than the highest molecular weight fraction (15 kDa and higher; likely contaminated by lignin-carbohydrate complexes), the other lignin fractions from different membranes have similar molecular structures (i.e., types of linkages among monolignols) despite their differences in molar mass [35]. Keyoumu et al. studied the ultrafiltration of hardwood (birch) and softwood (spruce-pine mixture) black liquor using ceramic membranes, and reported a negative correlation between membrane cut-offs and the phenolic group content in the obtained fractions [36].

5. Summary

Lignin fractionation has the potential of transforming low quality, heterogenous lignins to streams of lignin fractions with less impurities and narrower internal variation in chemical and structural properties, thus may enable the development of novel value-added products from lignin. To achieve this goal, cost-effective technologies that deliver lignin fractions with specific properties (e.g., low polydispersity, low ash, high/low phenolics content) are required. Methods for characterization of lignin fractions, as well as technologies for controlling lignin fraction characteristics, are instrumental in lignin fractionation for valorization. Therefore, new delignification, fractionation, characterization, and upgrading methods should be developed in tandem so that a complete pipeline of lignin valorization can be accomplished.

Author details

Zhenglun Li
Oregon State University, Corvallis, USA

*Address all correspondence to: glen.li@oregonstate.edu

References

[1] Humpert D, Ebrahimi M, Czermak P. Membrane Technology for the Recovery of lignin: A review. Membranes. 2016;**6**(3):42

[2] Menezes F, Rocha G, Filho RM. Obtainment and characterization of lignin from enzymatic hydrolysis of sugarcane bagasse of e2g process in pilot scale. Chemical Engineering Transactions. 2016;**50**:397-402

[3] Shevchenko SM, Beatson RP, Saddler JN. The nature of lignin from steam explosion/ enzymatic hydrolysis of softwood: Structural features and possible uses. Applied Biochemistry and Biotechnology. 1999;**79**(1-3):867-876

[4] Wang S, Bai J, Innocent MT, Wang Q, Xiang H, Tang J, et al. Lignin-based carbon fibers: Formation, modification and potential applications. Green Energy & Environment. 2022;**7**(4):578-605

[5] Furimsky E. Deactivation of hydroprocessing catalysts. Catalysis Today. 1999;**52**(4):381-495

[6] Lourençon TV, Hansel FA, da Silva TA, Ramos LP, de Muniz GIB, Magalhães WLE. Hardwood and softwood Kraft lignins fractionation by simple sequential acid precipitation. Separation and Purification Technology. 2015;**154**:82-88

[7] Helander M, Theliander H, Lawoko M, Henriksson G, Zhang L, Lindström ME. Fractionation of technical lignin: Molecular mass and pH effects. BioResources. 2013;**8**(2):2270-2282

[8] Norgren M, Edlund H, Wågberg L. Aggregation of lignin derivatives under alkaline conditions. Kinetics and Aggregate Structure. Langmuir. 2002;**18**(7):2859-2865

[9] Sewring T, Theliander H. Acid precipitation of Kraft lignin from aqueous solutions: The influence of anionic specificity and concentration level of the salt. Holzforschung. 2019;**73**(10):937-945

[10] Zhu W, Theliander H. Precipitation of lignin from softwood black liquor: An investigation of the equilibrium and molecular properties of lignin. BioResources. 2015;**10**(1):1696-1714

[11] Pan X, Kadla JF, Ehara K, Gilkes N, Saddler JN. Organosolv ethanol lignin from hybrid poplar as a radical scavenger: Relationship between lignin structure, extraction conditions, and antioxidant activity. Journal of Agricultural and Food Chemistry. 2006;**54**(16):5806-5813

[12] Gilarranz MA, Rodríguez F, Oliet M. Lignin behavior during the Autocatalyzed methanol pulping of Eucalyptus globulus changes in molecular weight and functionality. Holzforschung. 2000;**54**(4):373-380

[13] Li H, McDonald AG. Fractionation and characterization of industrial lignins. Industrial Crops and Products. 2014;**62**:67-76

[14] Izaguirre N, Robles E, Llano-Ponte R, Labidi J, Erdocia X. Fine-tune of lignin properties by its fractionation with a sequential organic solvent extraction. Industrial Crops and Products. 2022;**175**:114251

[15] Laurichesse S, Huillet C, Avérous L. Original polyols based on organosolv lignin and fatty acids: New bio-based building blocks for segmented

polyurethane synthesis. Green Chemistry. 2014;**16**(8):3958-3970

[16] Teramoto Y, Lee SH, Endo T. Molecular composite of lignin: Miscibility and complex formation of organosolv lignin and its acetates with synthetic polymers containing vinyl pyrrolidone and/or vinyl acetate units. Journal of Applied Polymer Science. 2012;**125**(3):2063-2070

[17] Fox S, McDonald A. Chemical and thermal characterization of three industrial lignins and their corresponding lignin esters. BioResources. 2010;**5**(2):990-1009

[18] Pan X, Saddler JN. Effect of replacing polyol by organosolv and Kraft lignin on the property and structure of rigid polyurethane foam. Biotechnology for Biofuels. 2013;**6**(1):12

[19] Evtuguin DV, Andreolety JP, Gandini A. Polyurethanes based on oxygen-organosolv lignin. European Polymer Journal. 1998;**34**(8):1163-1169

[20] Baker DA, Gallego NC, Baker FS. On the characterization and spinning of an organic-purified lignin toward the manufacture of low-cost carbon fiber. Journal of Applied Polymer Science. 2012;**124**(1):227-234

[21] Hosseinaei O, Harper D, Bozell J, Rials T. Improving processing and performance of pure lignin carbon Fibers through hardwood and herbaceous lignin blends. International Journal of Molecular Sciences. 2017;**18**(7):1410

[22] Gellerstedt G, Sjöholm E, Brodin I. The wood-based biorefinery: A source of carbon Fiber? The Open Agriculture Journal. 2010;**4**(1):119-124

[23] Klett AS, Chappell PV, Thies MC. Recovering ultraclean lignins of controlled molecular weight from Kraft black-liquor lignins. Chemical Communications. 2015;**51**(64):12855-12858

[24] Klett AS, Payne AM, Phongpreecha T, Hodge DB, Thies MC. Benign fractionation of lignin with CO_2 -expanded solvents of acetic acid + water. Industrial and Engineering Chemistry Research. 2017;**56**(34):9778-9782

[25] Chen Z, Ragauskas A, Wan C. Lignin extraction and upgrading using deep eutectic solvents. Industrial Crops and Products. 2020;**147**:112241

[26] Hou XD, Lin KP, Li AL, Yang LM, Fu MH. Effect of constituents molar ratios of deep eutectic solvents on rice straw fractionation efficiency and the micro-mechanism investigation. Industrial Crops and Products. 2018;**120**:322-329

[27] Li WX, Xiao WZ, Yang YQ, Wang Q, Chen X, Xiao LP, et al. Insights into bamboo delignification with acidic deep eutectic solvents pretreatment for enhanced lignin fractionation and valorization. Industrial Crops and Products. 2021;**170**:113692

[28] Yan G, Zhou Y, Zhao L, Wang W, Yang Y, Zhao X, et al. Recycling of deep eutectic solvent for sustainable and efficient pretreatment of corncob. Industrial Crops and Products. 2022;**183**:115005

[29] Liu Y, Chen W, Xia Q, Guo B, Wang Q, Liu S, et al. Efficient cleavage of lignin-carbohydrate complexes and ultrafast extraction of lignin oligomers from wood biomass by microwave-assisted treatment with deep eutectic solvent. ChemSusChem. 2017;**10**(8):1692-1700

[30] Piazza GJ, Lora JH, Garcia RA. Flocculation of high purity wheat straw

soda lignin. Bioresource Technology. 2014;**152**:548-551

[31] Abdelaziz OY, Ravi K, Nöbel M, Tunå P, Turner C, Hulteberg CP. Membrane filtration of alkali-depolymerised Kraft lignin for biological conversion. Bioresource Technology Reports. 2019;7:100250

[32] Feist CF, Hegeman GD. Phenol and benzoate metabolism by *Pseudomonas putida:* Regulation of tangential pathways. Journal of Bacteriology. 1969;**100**(2):869-877

[33] Linger JG, Vardon DR, Guarnieri MT, Karp EM, Hunsinger GB, Franden MA, et al. Lignin valorization through integrated biological funneling and chemical catalysis. Proceedings of the National Academy of Sciences. 2014;**111**(33):12013-12018

[34] Mendes SF, Pasquini D, Cardoso VL, Reis MHM. Ultrafiltration process for lignin-lean black liquor treatment at different acid conditions. Separation Science and Technology. 2022;**57**(12):1936-1947

[35] Toledano A, García A, Mondragon I, Labidi J. Lignin separation and fractionation by ultrafiltration. Separation and Purification Technology. 2010;**71**(1):38-43

[36] Keyoumu A, Sjödahl R, Henriksson G, Ek M, Gellerstedt G, Lindström ME. Continuous nano- and ultra-filtration of Kraft pulping black liquor with ceramic filters: A method for lowering the load on the recovery boiler while generating valuable side-products. 6th Int Lignin Inst Conf. 2004;**20**(2):143-150

Chapter 6

Perspective Chapter: Flaxseed (*Linum usitatissimum* L) – Chemical Structure and Health-Related Functions

Mehran Rahimlou and Jalal Hejazi

Abstract

Flaxseed (*Linum usitatissimum* L.), one of the oldest cultivated crops, continues to be widely grown for oil, fiber, and food. This herb is used in the food culture of some countries with different purposes. Flaxseed is one of the richest plants in terms of some nutrients. Numerous studies have shown that in terms of omega-3 fatty acids, this herb is one of the richest sources in the diet. Also, the lignan content in flaxseed is significantly higher than that in other plants. Also, the lignan content in flaxseed is significantly higher than that in other plants. In this chapter, the structure and chemical properties of this plant have been investigated first, and then the effectiveness of flaxseed in some chronic diseases has been mentioned.

Keywords: flaxseed, Lignan, omega-3, obesity, antioxidant, chemical composition

1. Introduction

Modern living presents a challenge to maintaining good health due to the prevalence of various chronic lifestyle disorders. Medical herbs have received the well-deserved attention as a result of the rapidly shifting international health situation and the quick understanding of the negative consequences of improper food preparation and overmedication. Strong demand in foods that can act as medicines is a result of growing understanding of the importance of diet and the pursuit of well-being. Foods or dietary components known as "functional foods" or "neutraceuticals" can also provide health benefits above and beyond those of simple nutrition. Beyond what might be predicted from their typical nutritive value, functional foods improve health. One of the useful compounds that has attracted a lot of attention from researchers in recent years is flaxseed.

Flaxseed (*Linum usitatissimum* L.) is one of the predominant industrial oils seed crops grown in temperate climates. The terms flaxseed (*Linum usitatissimum* L.) and linseed are often used interchangeably. The seed oil of this oilseed crop is enriched in α-linolenic acid (ALA) (18:3cisΔ9,12,15). In addition to linolenic and linoleic acids, lignans, cyclic peptides, polysaccharides, alkaloids, cyanogenic glycosides, and cadmium are among the numerous physiologically active substances and elements that concentrate in flaxseed. The majority of a-linolenic acid or lignan-containing extracts have been the subject of biological and clinical studies on flaxseed. In this chapter, we explain about the chemical characteristics and clinical effects of flaxseed in improving health and preventing chronic diseases.

2. Origin and domestication

Flax, whose scientific name is *Linum usitatissimum* L., which translates to "extremely beneficial," comes in two primary colors: brown and yellow or golden (also known as golden linseeds) [1]. The genus Linum is a large group with ~230 species, which subclassed to fivesubgroups including: depending on the number of chromosomes, the shape of flowers, and interspecific adaptability [2]. The section Linum contains the cultivated flax, *Linum usitatissimum*, which has 30 diploid chromosomes, along with the proposed progenitors *L. angustifolium* and L.bienne, which share homostylous rather than heterosylous flowers [3, 4].

The seed of the annual herb known as the flax plant, a member of the Linaceae family, is known as flaxseed. This herb is planted in different regions of the world from eastern Mediterranean to India, Western Asia, and the Middle East, and today, it accounts for a part of the global trade of agricultural products of some countries. There is documentation that flax was used as a fiber source in Neolithic societies. Evidence of flax fiber has been found in ancient sites in Israel and Syria [5]. Between 4500 and 4000 BC, flax was farmed in Egypt. It then spread across Switzerland and was discovered much later at Willium Hill in England [6]. Flax is thought to have been introduced to Canada by Lois Hebert, the country's first farmer, over 400 years ago [7]. Nowadays, the term "flaxseed" refers to the seeds of the flax plant when they are eaten by human, whereas the phrase "linseed" refers to the seeds of the flax plant when they are used for industrial and feed purposes [8]. The usual life cycle of a flax plant is 45–60 days of vegetative time, 15–25 days of flowering time, and 30–40 days of maturity time. It can reach a height of 12–36 meters. In accordance with the cultivar, flax can produce blue, white, pink, or violet flowers. The flax plant's mature fruit is a five-segmented dry boll or capsule. Two seeds are contained in each segment of the boll; hence, a capsule can contain up to 10 (on average 6–8) flaxseeds [9]. Seeds of flax appear in different colors depending on the amount of tannin pigments in the pigment cells of seed coat and included medium, reddish brown, and light yellow, where the brown seeds are the most abundant form of flaxseeds with higher tannin contents [10]. A commercial flaxseed strain with yellow seeds called Solin (Linola) was introduced and sold in Canada [11].

Since its cultivation began in Mesopotamia about 5000 BC, flaxseed has mostly been used for the production of cloth and paper. As mentioned, the global trade of this plant today accounts for a part of the global economy of agricultural products. Canada is the largest producer of this plant in the world with an annual production of 20 tons, and after this country, other countries such as America, China, India, and Russia are also among the main producers of this plant in the world [12]. As mentioned, the global trade of this plant today accounts for a part of the global economy of agricultural products. Canada and Russia are among the largest producers of this plant in the world, and after this country, other countries such as America, China, India, and Kazakhstan are also among the main producers of this plant in the world [13, 14].

3. Flaxseed structure and chemical composition

An embryo with two cotyledons that is enclosed by a thin endosperm and a smooth, frequently lustrous yellow to dark brown seed coat makes up mature flaxseed, which is oblong and flattened (hull) [15]. Flaxseed sprouts are edible, with a slightly spicy flavor. Whole flax seeds are chemically stable, but ground flaxseed can go rancid at room temperature in as little as 1 week, although there is contrary evidence. A longer amount of time will be prevented from ground flax going rancid by refrigeration and storing in airtight containers. When packed quickly without exposure to air or light, milled flax is impressively resilient to oxidation for 9 months at room temperature and for 20 months at ambient temperatures under warehouse conditions [16, 17].

The components of this herb are summarized in **Table 1**. Also, the content of vitamins, minerals, and active compounds in flaxseed is summarized in **Table 2**. The protein content in flaxseed has been reported as 10–30% in some studies [18–20]. In terms of protein distribution, the highest amount of protein is in the cotyledons of the flaxseed (between 50 and 70%), and about 30% is in the coat and endosperm [21]. The amino acid profile of flaxseed protein is comparable with that of soy. As shown in **Table 1**, higher amounts of the some amino specially arginine, valine, glycine, leucine, valine, and serine have been found in flaxseed [22]. In terms of essential amino acid index of flaxseed, it has been reported that the score for this index for flaxseed is 69, and this score is near the score for canola and soy, that is, 75 and 79, respectively [23]. Also, according to the Food and Agriculture Organization (FAO) reports, flaxseed score based on the limiting amino acid is 82; however, this score for soybean is 67 [24]. It has also been shown in some studies that 34.3% of the amino acids in flaxseed are essential amino acids, and the ratio of lysine to arginine, which is an indicator of cholesterolemic and atherogenic effects of a protein, is about 0.22–0.37, which reveals that flaxseed is less atherogenic than canola and soybean [25].

The Canadian Grain Commission examined brown Canadian flaxseed, finding that it typically included 41% fat, 20% protein, 28% total dietary fiber, 7.7% moisture, and 3.4% ash [3]. As reported, flaxseeds are rich in oil (40–46% by seed weight) with a healthy fatty acid profile. In terms of type of fatty acids, flaxseed contains about 73% polyunsaturated fatty acids (PUFAs) and 10%

Components (%)	Whole flaxseed	Component	g/100 g of flaxseed	Component	g/100 g of flaxseed
Moisture	7.13	α-linolenic acid	22.8	Arginine	9.2
Nitrogen	4.01	Linoleic acid	5.9	Glycine	5.8
Fat	41	Oleic acid	7.3	Leucine	5.8
Protein	20	Stearic acid	1.3	Valine	4.6
Total dietary fiber	28	Palmitic acid	1.2	Serine	4.5
Ash	3.4	Glutamic acid	19.6	Soluble Fibers	4.3–8.6
		Aspartic acid	9.3	Insoluble Fibers	12.8–17.1

Table 1.
The most important flaxseed components.

Components	g/100 g of flaxseed	Component	g/100 g of flaxseed	Phenolic Compounds mg/100 g of flaxseed	
Calcium	236	δ-tocopherol	10	Secoisolariciresinol	165
Magnesium	431	vitamin C	0.5	Laricinesol	1.7
Phosphorus	622	vitamin B1	0.5	Pinoresinol	0.8
Potassium	831	vitamin B2	0.2	Total Flavonoids	35–70
Sodium	27	vitamin B3	1.2		
Zinc	4	vitamin B6	0.6		
Iron	5	Pantothenic acid	0.6		
Manganese	3	**Phenolic Compounds mg/g flaxseed powder**			
γ-tocopherol	522	Ferulic acid	10.9		
α-tocopherol	7	Chlorogenic acid	7.5		
		Gallic acid	2.8		

Table 2.
The content of vitamins, minerals, and active compounds in flaxseed.

saturated fatty acids. Also, in terms of the type of PUFAs, the evaluation results on different samples of flaxseed show the presence of about 52–57% α-linolenic acid (ALA) and an ω-3/ω-6 ratio of 1:0.3 [26]. Considering that higher amounts of ALA make the oil prone to oxidation and Linola seeds have about 5% less ALA compared with the brown flaxseed, therefore, this type of flaxseed is more resistant to oxidation than brown flaxseed and for applications such as frying and shortenings are more suitable [11].

As shown in **Table 2**, in terms of minerals and vitamins content, flaxseed is one of the appropriate sources of calcium, phosphorus, potassium, magnesium, and γ-tocopherol [3]. In addition, flaxseed is a good source of some active compounds such as phenolic compounds. It has been reported that some of the bioactive phenolic compounds are ferulic acid, chlorogenic acid, and gallic acid [27]. Cyanogenic glycosides (CG), phytates, lignans, and antipyridoxin factors are some of flaxseed's minor constituents that have been identified. Flaxseed has amounts of the lignan secoisolariciresinoldiglucoside (SDG) that are 75–800 times higher than those of any other documented plants or vegetables [26]. Flaxseed is one of the best dietary functional foods in terms of supply lignan, and the amount of lignan in it varies between 0.9 and 3% depending on various factors [11]. Based on the evidence in some experimental studies, the lignans in flaxseed have anti-estrogenic effects and may have a preventive role in the pathological process of some hormone-dependent cancers [28]. The most predominant lignan in flaxseed is SDG. Enterolactone and enterodiol, which have antiestrogenic properties and structural similarities to estrogen and can bind to cell receptors to inhibit cell proliferation, are formed from SDG [28]. Numerous items made from flaxseed are being sold in the market and promoted for their benefits as functional foods and nutraceuticals. Whole seeds, ground or milled flaxseed (flour), oil extracted from flaxseed (by pressing; cold-pressed or not), flax meal, the coat of the seed, a portion of the seed that has been removed, flaxseed hull, cyclic peptides (orbitides) from flaxseed oil, and lignan extracts from the flaxseed hull are among them [29]. Some of the experimental studies evaluated the digestibility of proteins in the flaxseed, and in some studies, the digestibility percent of this plant has been reported as 81.4–85.8% [30, 31], and this is near the soybean digestibility percent (digestibility percent = 84–85%) [32]. Also, the flaxseed biological value (BV) ranges from 66.4% to 77.4% [33]. The matrix elements, particularly mucilage and oil, and the earlier preparation of the seeds have an impact on the flaxseed protein digestibility [34]. It has been reported in some experimental studies that elimination of flaxseed mucilage improved protein digestibility to 50% compared with the 12.6% for full-fat and mucilage containing ground flaxseed [35]. Also, it has been reported that some of the other processing methods such as boiling, heat treatment can improve flaxseed protein digestibility [36]. One of the concerns related to flaxseed is allergy to the its proteins. However, to date, the results of several studies have reported rare cases of allergy to the proteins contained in flaxseed. Some researchers believe that this allergy observed in some cases is not due to flaxseed proteins, but the reason is the contamination in some seeds [37].

4. Flax: a functional food

It is appropriate to classify flax as a "functional food" due to the numerous health-promoting components found in both the seeds and the oil. Foods known as "functional foods" have potential health benefits in addition to its nutrient contents. L-arginine, cyanogenic glycosides (CNglcs), and omega-3 fatty acids can be mentioned among the active components in the seeds that cause its positive functions in the prevention and treatment of some chronic diseases. The consumer can currently choose between whole flaxseeds and milled flaxseed in the area of health-related functional foods. These goods are promoted as excellent providers of lignans, dietary fiber, and omega-3 fatty acids. Numerous common foods, including muffins, bread, crackers, tortillas, cereals, and snack bars, include whole flaxseeds. The commercial marketing of flaxseed products as sources of protein for food usage is extremely limited [38].

However, in recent years, some food manufacturing companies use flaxseed to create a pleasant nutty flavor in the preparing cereal and bakery products. Also, due to the fact that flaxseed does not contain gluten, some food companies use flax flour in the production of gluten-free food for celiac patients [39]. Some attempts have been made to fortify breads with flaxseed flour; however, some limitations have made the results not very satisfactory. For example, it has been reported that use of flaxseed flour for bread production resulted in poor texture, crumb color, graininess, volume, and crust color [40]. On the other hand, considering that flaxseed contains PUFA, the use of this plant in the bakery industry and exposure to high heat may cause the oxidation of fatty acids and the production of metabolites such as malondialdehyde (MDA), which have adverse effects on health [41]. However, Cunnane et al. were reported that cooking at the normal temperature and for a limited time does not significantly increase the production of MDA [42]. Antinutritional components included in flaxseed products present a significant obstacle to their use in food products. Linustatin, neolinustatin, and linamarin are CGs that can be found in flaxseed [43]. These substances are broken down by β-glucosidases when the cell structure is disturbed, generating small amounts of hydrogen cyanide, a potent respiratory inhibitor [44]. However, in some studies, it has been reported that usage from some methods such as extraction of flaxseed meal 2–3 times with methanol-ammonia-water, heat treatment, boiling for 5 min in water, and dry and wet autoclaving can decrease CG in flaxseed [45, 46]. Linatine, an antagonist of vitamin B6 (pyridoxine), is also found in flaxseed. Generally, linatine concentrations range from 20 to 100 mg/kg. Before eating, heat processing could be required to eliminate these antinutritional elements [47].

5. Extraction methods

There are a number of methods that have been suggested for extracting oil from flaxseed, most notably mechanical extraction, solvent extraction, supercritical/subcritical fluid extraction, and extraction with microwave or ultrasonic intervention. The constitution and amount of unsaturated fatty acids and active substances extracted using various techniques typically vary significantly due to the complexity of flaxseed oil. Flaxseed oil is traditionally produced through mechanical pressing, which depends on outside machinery to produce oil. Among the advantages of using mechanical methods to extract flaxseed oil can be mentioned the low cost and equipment, no change in the taste of the oil, acceptable quality, and no need to use chemical compounds for the extraction process [48]. It has been reported that using from the double-screw extrusion preserves the flaxseed oil quality and its ALA contents, but there are also some clear disadvantages, including low yield, high residual oil, and significant reduction in concentration of some of vitamin, phospholipids, phytosterols, and antioxidants [49]. Low extraction efficiency is properly remedied by solvent method, and up to 99% of the oil is obtained. However, there are some concerns that the use of acetone, methanol, petroleum ether, n-hexane dichloromethane, ethanol, and heptane may decrease the purity of oil, influence its physiological activities, and even result in potential issues with food safety and environmental contamination. As a result, instead of producing commercial products, this technology is primarily employed for laboratory-size investigation.

Modern methods of oil extraction have received a lot of attention recently, including supercritical CO_2 extraction, microwave-assisted extraction, ultrasound-assisted extraction, and subcritical fluid extraction. Nowadays, one of the modern methods suggested for extracting flaxseed oil is the use of supercritical CO_2 fluid extraction

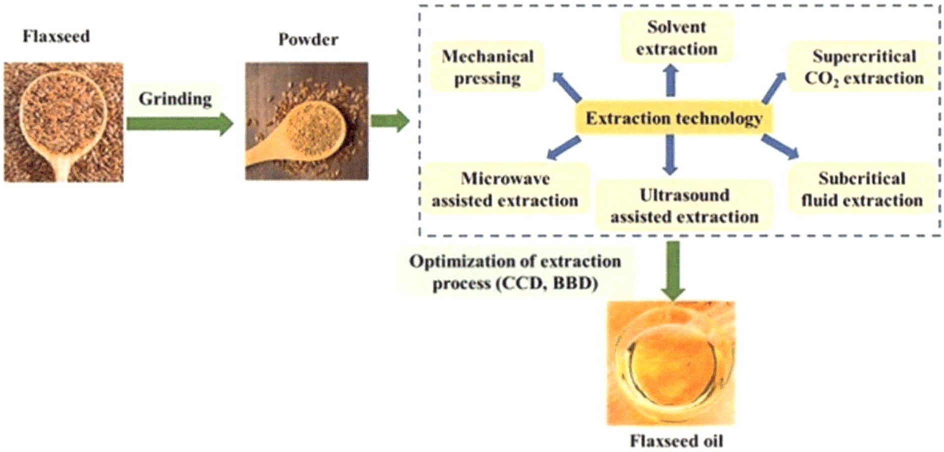

Figure 1.
Representative extraction methods of flaxseed oil [53].

method, which, in addition to the high quality of the oil, preserves the high concentration of omega-3 fatty acids, phenolic compounds, and lignans [50], and the production efficiency is about 25% higher than that in the mechanical method [51]. Using the microwave and ultrasonic-assisted extraction methods compared with the solvent method has significant advantages, including shortening the extraction time, reducing the solvent consumption, thermal damage to the extract, and the loss of bioactive compounds [52]. The commonly used extraction methods are shown in **Figure 1**.

6. Biological effects of flaxseed and flaxseed fractions

Whole flaxseed, ground flaxseed, flaxseed oil, partially defatted flaxseed meal (typically from expeller pressing), fully defatted flaxseed meal (from solvent extraction), flaxseed mucilage extracts, flaxseed hulls, flaxseed oleosomes, and flaxseed alcohol extracts are some of the flaxseed products that are consumed. These substances each have unique positive health impacts that have been linked to them. In the following, we discuss some of the positive biological effects of flaxseed in improving health and preventing and treating chronic diseases.

7. Central activities

The higher ALA concentration of flaxseed has been linked to the majority of its effects. Three distinct metabolic destinations are possible for ALA: (a) absorption into structural, transport, or storage pools; (b) use as an energy source via beta-oxidation; and (c) extension and further desaturation to produce EPA, DPA, and DHA.

Although alpha-linolenic acid plays a small role in the cell membrane directly, it plays an important role indirectly in the cell structure after being converted into long-chain omega-3 fatty acids such as EPA and DHA [54]. On the other hand, a part of ALA consumed through diet participates in energy production processes in the body through beta oxidation, the results of some studies have shown that the percentage of ALA that participates in beta oxidation is about 33% in menand 22% in women [55]. On the other hand, the ALA present in flaxseed reduces the production of inflammatory eicosanoids through metabolic pathways [56]. Additionally, some of the activities of flaxseed may not be reliant on its fatty acid concentration and type and may instead be caused by the lignan SDG. It has been established that the SDG is the primary lignin in flaxseed, and that the gastrointestinal bacteria convert this lignin into the mammalian lignans enterodiol (ED) and enterolactone (EL) [57].

8. Anti-obesity effects of flaxseed

Obesity is a serious health issue and a significant contributor to the risk of diabetes, heart disease, and cancer. By 2030, it is anticipated that 1.12 billion people worldwide would be obese [58]. Type 2 diabetes, hypertension, inflammation, metabolic syndrome, cardiovascular dysfunction, and non-alcoholic fatty liver disease are among the comorbid conditions that the majority of obese persons have [59]. Also, it has been reported in several studies that obesity is an important risk factor for some type of cancers [60]. As a result, preventing obesity is essential for lowering chronic disease rates. Several studies have evaluated the effect of flaxseed supplementation in obese people.

Mohammadi-Sartang in a meta-analysis from 45 randomized controlled trial studies and 2789 participants showed that flaxseed consumption led to a significant reduction in body weight, body mass index, and waist circumference. Also, they found that the flaxseed supplementation caused significant weight loss if flaxseed was taken with a dose higher than 30 grams per day, and the duration of the intervention was more than 12 weeks [61]. Several mechanisms have been proposed for the anti-obesity effects of flaxseed. As mentioned, flaxseed is one of the richest sources of lignan (primarily SDG), and its lignan content is up to 100 times higher than that of some other plants [62, 63]. SDG, a type of purified lignans, has been demonstrated to decrease visceral (abdominal) fat in mice when compared with animals that did not consume SDG [64]. The researchers proposed that a drop in the mRNA levels of sterol regulatory element-binding proteins, which are essential for the synthesis of triacylglycerol, may result from flaxseed's high amount of SDG [64]. Also, the results of some studies have shown that the lignans in flaxseed regulate the expression of adiponectin and stimulate fat oxidation, all of which prevent excessive fat accumulation and obesity [64, 65]. It has been demonstrated that dietary fiber, especially soluble fiber, has anti-obesity properties that reduce body weight. By delaying gastric emptying, causing sensations of fullness via absorbing significant amounts of water, and/or increasing the levels of short chain fatty acids, which operate to induce satiety by a wide range of mechanisms, dietary fiber can help control weight or aid weight loss [66].

9. Anti-inflammatory action

Inflammation is one of the main factors in the pathology of many chronic diseases. An increase in the level of inflammatory cytokines in the body provides the basis for damage to various tissues. In some studies, it has been shown that flaxseed has strong anti-inflammatory properties due to its many active compounds, especially because of its ALA contents [67]. The results of a meta-analysis study designed and implemented by our team in 2018 showed that supplementing with flaxseed had a significant effect on reducing the concentration of some inflammatory factors including tumor necrosis factor alpha and high-sensitivity C-reactive protein (hs-CRP) [68]. Positive results have also been reported from supplementing with flaxseed in inflammatory diseases such as rheumatoid arthritis [69] and inflammatory bowel diseases [70]. SDG and its metabolites (enterolactone and lignans enterodiol), which are flaxseed-related active chemicals, in particular, have demonstrated anti-inflammatory and antioxidant potential, primarily by inhibiting lipid peroxidation. Additionally, some research has indicated that the flaxseed lignan portion may stimulate the nuclear element (erythroid-derived 2)-like 2, a transcription factor of the antioxidant and detoxifying genes, including NAD(P)H quinone dehydrogenase 1 and heme oxygenase-1 [71]. It's important to take into account the connections between the microbiome, inflammation, and diet. For instance, flaxseed is the greatest non-bioactive supply of lignan precursors and is transformed into its main metabolites, which have antioxidant activity, in the colon by gut microbiota, which may be altered by the microbial diversity [72]. On the other hand, flaxseed, as one of the richest plant sources of omega-3, can exert its anti-inflammatory effects through the conversion of alpha-linolenic acid into long-chain omega-3 fatty acids such as eicosapentaenoic acid (EPA) and Docosahexaenoic acid (DHA). The production of pro-inflammatory eicosanoids from omega-6 fatty acids, such as prostaglandin E2 (PGE2) and leukotriene B4 (LTB4), suppressed by omega 3 fatty acids and can induce the production of some anti-inflammatory agents such as resolvins and protectins. Together these provide a mechanism to suppress the activity of nuclear transcription factors including nuclear factor kappa B (NF-κB) and thus decrease the production of pro-inflammatoryenzymes and cytokines, such as cyclooxygenase-2(COX-2), TNF-α, and interleukin (IL)-1 [71, 73].

10. Cardiovascular health

Cardiovascular diseases are one of the main causes of death in the world and one of the five main causes of death and disability in most countries. About 2200 Americans every day, or one every 40 seconds, pass away from cardiovascular disease in the United States [74]. Numerous clinical studies, including DART [75], the GISSI Prevenzione trial [76], GISSI-HF [77], have declared that omega 3 supplementation led to a significant cardioprotective effects. Several studies have investigated the cardioprotective effects of flaxseed from different aspects. This includes dietary supplementation with flaxseed and its effects on atherogenesis [78], lipid profile [79], platelet aggregation [80], arrhythmias [81], hypertension [82], and oxidative stress [83]. In a clinical trial on 112 patients with high blood pressure, Toulabi et al. showed that supplementation with 30 grams of flaxseed per day caused a significant decrease in systolic anddiastolic blood pressure [84]. Also, Mahmudiono et al. in a meta-analysis study showed that flaxseed oil supplementation among the patients with metabolic syndrome and related disorders caused a significant reduction in systolic but not diastolic blood pressure [85]. It is unclear how flaxseed lowers blood pressure and how this happens. Weight loss has been advocated as an effective method for treating high blood pressure in overweight and obese people [56]. But according to one study, flaxseed contains peptides that are high in arginine and can lower blood pressure [86]. Furthermore, flaxseed has a high concentration of ALA. It has been extensively evaluated in some studies and shown that ALA can reduce blood pressure and reduce cardiovascular diseases risk [87, 88].

Also, several studies have evaluated the effect of flaxseed on the lipid profile. Hadi et al. in an updated systematic review and dose-response meta-analysis of 62 randomized controlled trials evaluated the effects of flaxseed on lipid profile, and it has been reported that flaxseed supplementation significantly reduced total cholesterol (TC), triglyceride (TG), and low-density lipoprotein cholesterol (LDL-C). However, there were not any significant effects on high-density lipoprotein (HDL-C). Also, they found that flaxseed oil had no effect compared with whole flaxseed or lignin [89]. As mentioned, these beneficial dyslipidemia effects are caused by the content of lignan, phenolic, and ALA in flaxseed. By boosting satiety, lowering caloric intake, and decreasing meal transit time, flaxseed's high fiber content protects against dyslipidemia. It also stimulates bile acid excretion and reduces bile acid reabsorption through enhanced fecal excretion of cholesterol. Additionally, fiber affects other metabolic processes, such as fatty acid biosynthesis, lipolysis, and cholesterol synthesis by forming short-chain fatty acids that prevent the formation of liver cholesterol [90].

11. Conclusion

Flaxseed is one of the richest herbs in terms of nutrients and active compounds. This herb is known as one of the best dietary sources of omega-3 fatty acids and lignan.

This herb is used in different forms, the most common of which are the whole flaxseed, ground flaxseed, and flaxseed oil. Several methods are used to extract flaxseed oil from full flaxseed, from mechanical methods that are part of traditional methods to more modern methods such as microwave and ultrasonic-assisted extraction methods. In recent years, several studies have been conducted on the beneficial properties of flaxseed in the prevention and treatment of chronic diseases as an alternative treatment, which include anti-inflammatory, antioxidant, anti-obesity effects, improved lipid profile, and reduced risk of cardiovascular diseases. Part of these beneficial effects of this plant is due to its high content of lignans. Despite the mentioned beneficial effects, there are also some concerns related to the use of this plant, such as the risk of oxidation of flaxseed oil in higher temperature due to higher content of PUFA and flaxseed contamination with toxins and heavy metals. Also, some researchers have reported concerns regarding the use of this plant in people with hormone-dependent cancers such as breast and ovarian cancer due to its phytoestrogens. Therefore, there is a need for more studies related to the different aspects of using this plant in the daily diet as well as in the prevention and treatment of chronic diseases.

Acknowledgements

The authors thank the Zanjan University of Medical Sciences for their support in preparing resources and articles for writing this chapter of the book. Both authors (MR and JH) contributed equally in writing all parts of this chapter.

Conflict of interest

The authors declare that there was no conflict of interest in the writing of this chapter.

Author details

Mehran Rahimlou* and Jalal Hejazi
Department of Nutrition, School of Medicine, Zanjan University of Medical Sciences, Zanjan, Iran

*Address all correspondence to: rahimlum@gmail.com

References

[1] Daun J et al. Structure, composition, and variety development of flaxseed. Flaxseed in human nutrition. 2003;**35**(2):1-40

[2] Campos JR, Severino P, Ferreira CS, Zielinska A, Santini A, Souto SB, et al. Linseed essential oil–source of lipids as active ingredients for pharmaceuticals and nutraceuticals. Current Medicinal Chemistry. 2019;**26**(24):4537-4558

[3] Shim YY et al. Flaxseed (*Linum usitatissimum* L.) bioactive compounds and peptide nomenclature: A review. Trends in Food Science & Technology. 2014;**38**(1):5-20

[4] Ramesh M. Flax (*Linum usitatissimum* L.) fibre reinforced polymer composite materials: A review on preparation, properties and prospects. Progress in Materials Science. 2019;**102**:109-166

[5] Van Zeist W, Bakker-Heeres JA. Evidence for linseed cultivation before 6000 BC. Journal of Archaeological Science. 1975;**2**(3):215-219

[6] Smith G et al. A study of rope workers exposed to hemp and flax. Occupational and Environmental Medicine. 1969;**26**(2):109-114

[7] Hall LM et al. Flax (*Linum usitatissimum* L.). In: Industrial oil crops. UK: Elsevier; 2016. pp. 157-194

[8] Morris DH. Flax: A Health and Nutrition Primer. Canada: Flax Council of Canada; 2007

[9] Mueed A, Shibli S, Korma SA, Madjirebaye P, Esatbeyoglu T, Deng Z. Flaxseed Bioactive Compounds: Chemical Composition, Functional Properties, Food Applications and Health Benefits-Related Gut Microbes. Foods. 2022;**11**(20):3307

[10] Diederichsen A, Richards K. Cultivated flax and the genus Linum L.: Taxonomy and germplasm conservation. In: flax. Canada: CRC Press; 2003. pp. 34-66

[11] Westcott ND, Muir AD. Flax seed lignan in disease prevention and health promotion. Phytochemistry Reviews. 2003;**2**(3):401-417

[12] Oomah BD. Flaxseed as a functional food source. Journal of the Science of Food and Agriculture. 2001;**81**(9):889-894

[13] Morris DH. Essential nutrients and other functional compounds in flaxseed. Nutrition Today. 2001;**36**(3):159-162

[14] Raghuwanshi V, Agrawal R, Mane K. Flaxseed as a functional food: A review. J Pharmacogn Phytochem. 2019;**8**:352-354

[15] Ayash AA, Raouf RM, Ibrahim YK. An investigation of flaxseed effect on the mechanical behaviour of cement mortar. Materials Today: Proceedings. 2021;**42**:2616-2620

[16] Malcolmson L, Przybylski R, Daun J. Storage stability of milled flaxseed. Journal of the American Oil Chemists' Society. 2000;**77**(3):235-238

[17] Bernacchia R, Preti R, Vinci G. Chemical composition and health benefits of flaxseed. Austin Journal of Nutrition and Food Science. 2014;**2**(8):1045

[18] Oomah B, Mazza G. Flaxseed proteins—A review. Food Chemistry. 1993;**48**(2):109-114

[19] Khattab R, Arntfield S, Nyachoti C. Nutritional quality of legume seeds as affected by some physical treatments,

part 1: Protein quality evaluation. LWT-food Science and Technology. 2009;**42**(6):1107-1112

[20] Sultan A et al. Nutritional effect of flaxseeds on cholesterol profile and fatty acid composition in egg yolk. Cereal Chemistry. 2015;**92**(1):50-53

[21] Dev DK, Quensel E, Hansen R. Nitrogen extractability and buffer capacity of defatted linseed (*Linum usitatissimum* L.) flour. Journal of the Science of Food and Agriculture. 1986;**37**(2):199-205

[22] Bhatty R, Cherdkiatgumchai P. Compositional analysis of laboratory-prepared and commercial samples of linseed meal and of hull isolated from flax. Journal of the American Oil Chemists' Society. 1990;**67**(2):79-84

[23] Waszkowiak K, Siger A, Rudzińska M, Bamber W. Effect of roasting on flaxseed oil quality and stability. Journal of the American Oil Chemists' Society. 2020;**97**(6):637-649

[24] Sosulski F, Sarwar G. Amino acid composition of oilseed meals and protein isolates. Canadian Institute of Food Science and Technology Journal. 1973;**6**(1):1-5

[25] Marambe HK, Shand PJ, Wanasundara JP. Release of angiotensin I-converting enzyme inhibitory peptides from flaxseed (*Linum usitatissimum* L.) protein under simulated gastrointestinal digestion. Journal of Agricultural and Food Chemistry. 2011;**59**(17):9596-9604

[26] Madhusudhan B. Potential benefits of flaxseed in health and disease-a perspective. Agriculturae Conspectus Scientificus. 2009;**74**(2):67-72

[27] El-Beltagi H, Salama Z, El-Hariri D. Evaluation of fatty acids profile and the content of some secondary metabolites in seeds of different flax cultivars (*Linum usitatissimum* L.). general and applied. Plant Physiology. 2007;**33**(3-4):187-202

[28] Calado A et al. The effect of flaxseed in breast cancer: A literature review. Frontiers in Nutrition. 2018;**5**:4

[29] Shim YY et al. Flaxseed (*Linum usitatissimum* L.) oil processing and selected products. Trends in Food Science & Technology. 2015;**43**(2):162-177

[30] Hossain M, Jauncey K. Studies on the protein, energy and amino acid digestibility of fish meal, mustard oilcake, linseed and sesame meal for common carp (*Cyprinus carpio* L.). Aquaculture. 1989;**83**(1-2):59-72

[31] Hossain M, Nahar N, Kamal M. Nutrient digestibility coefficients of some plant and animal proteins for rohu (Labeo rohita). Aquaculture. 1997;**151**(1-4):37-45

[32] Hussain S et al. Chemical composition and functional properties of flaxseed (*Linum usitatissimum*) flour. Sarhad Journal of Agriculture. 2008;**24**(4):649-653

[33] Martinchik A, Baturin A, Zubtsov V, Molofeev V. Nutritional value and functional properties of flaxseed. Vopr Piton. 2012;**81**(3):4-10

[34] Marambe HK, Shand PJ, Wanasundara JP. In vitro digestibility of flaxseed (*Linum usitatissimum* L.) protein: Effect of seed mucilage, oil and thermal processing. International Journal of Food Science & Technology. 2013;**48**(3):628-635

[35] Marambe H, Wanasundara J. Protein from flaxseed (*Linum usitatissimum* L.).In: Sustainable protein sources. Canada: Elsevier; 2017. pp. 133-144

[36] Giacomino S et al. Extruded flaxseed meal enhances the nutritional quality of cereal-based products. Plant Foods for Human Nutrition. 2013;**68**(2):131-136

[37] Muir AD, Westcott ND. Flaxseed constituents and human health. In: Flax. Canada New York: CRC Press; 2003. pp. 255-263

[38] Goyal A et al. Flax and flaxseed oil: An ancient medicine & modern functional food. Journal of Food Science and Technology. 2014;**51**(9):1633-1653

[39] Gupta M, Bansal V. Development of gluten free snacks using chickpea flour and flax seeds for celiac patients. Journal of Emerging Technologies and Innovative Research. 2021;**8**:223-238

[40] Conforti FD, Davis SF. The effect of soya flour and flaxseed as a partial replacement for bread flour in yeast bread. International Journal of Food Science & Technology. 2006;**41**:95-101

[41] Alpers L, Sawyer-Morse MK. Eating quality of banana nut muffins and oatmeal cookies made with ground flaxseed. Journal of the Academy of Nutrition and Dietetics. 1996;**96**(8):794

[42] Cunnane SC et al. High α-linolenic acid flaxseed (*Linum usitatissimum*): Some nutritional properties in humans. British Journal of Nutrition. 1993;**69**(2):443-453

[43] Parikh M, Maddaford TG, Austria JA, Aliani M, Netticadan T, Pierce GN. Dietary flaxseed as a strategy for improving human health. Nutrients. 2019;**11**(5):1171

[44] Wanasundara P et al. Removal of cyanogenic glycosides of flaxseed meal. Food Chemistry. 1993;**48**(3):263-266

[45] Chadha R, Lawrence J, Ratnayake W. Ion chromatographic determination of cyanide released from flaxseed under autohydrolysis conditions. Food Additives & Contaminants. 1995;**12**(4):527-533

[46] Madhusudhan K, Singh N. Effect of heat treatment on the functional properties of linseed meal. Journal of Agricultural and Food Chemistry. 1985;**33**(6):1222-1226

[47] Chaudhary B et al. Uses of flax (*Linum usitatisimum*) after harvest. International Journal of Tropical Agriculture. 2016;**34**(4):159-164

[48] Bhargavi G, Rao PN, Renganathan S. Review on the extraction methods of crude oil from all generation biofuels in last few decades. In: IOP Conference Series: Materials Science and Engineering. China: IOP Publishing; 2018

[49] Kasote DM, Badhe YS, Hegde MV. Effect of mechanical press oil extraction processing on quality of linseed oil. Industrial Crops and Products. 2013;**42**:10-13

[50] Khattab RY, Zeitoun MA. Quality evaluation of flaxseed oil obtained by different extraction techniques. LWT-Food Science and Technology. 2013;**53**(1):338-345

[51] Pradhan RC et al. Supercritical CO2 extraction of fatty oil from flaxseed and comparison with screw press expression and solvent extraction processes. Journal of Food Engineering. 2010;**98**(4):393-397

[52] Tang Z-X, Ying R-F, Lv B-F, Yang L-H, Xu Z, Yan L-Q, et al. Flaxseed oil: Extraction, Health benefits and products. Quality Assurance and Safety of Crops & Foods. 2021;**13**(1):1-19

[53] Yang J et al. The composition, extraction, analysis, bioactivities, bioavailability and applications in food system of flaxseed (*Linum usitatissimum*

L.) oil: A review. Trends in Food Science & Technology. 2021;**118**:252-260

[54] Arterburn LM, Hall EB, Oken H. Distribution, interconversion, and dose response of n− 3 fatty acids in humans. The American Journal of Clinical Nutrition. 2006;**83**(6):1467S-1476S

[55] Burdge G. α-Linolenic acid metabolism in men and women: Nutritional and biological implications. Current Opinion in Clinical Nutrition & Metabolic Care. 2004;**7**(2):137-144

[56] Rezaei S et al. Flaxseed oil in the context of a weight loss programme ameliorates fatty liver grade in patients with non-alcoholic fatty liver disease: A randomised double-blind controlled trial. British Journal of Nutrition. 2020;**123**(9):994-1002

[57] Prasad K. Flaxseed and cardiovascular health. Journal of Cardiovascular Pharmacology. 2009;**54**(5):369-377

[58] Kelly T et al. Global burden of obesity in 2005 and projections to 2030. International Journal of Obesity. 2008;**32**(9):1431-1437

[59] Heindel JJ, Newbold R, Schug TT. Endocrine disruptors and obesity. Nature Reviews Endocrinology. 2015;**11**(11):653-661

[60] Avgerinos KI et al. Obesity and cancer risk: Emerging biological mechanisms and perspectives. Metabolism. 2019;**92**:121-135

[61] Mohammadi-Sartang M et al. The effect of flaxseed supplementation on body weight and body composition: A systematic review and meta-analysis of 45 randomized placebo-controlled trials. Obesity Reviews. 2017;**18**(9):1096-1107

[62] Muir AD. Flax lignansanalytical methods and how they influence our understanding of biological activity. Journal of AOAC International. 2006;**89**(4):1147-1157

[63] Bhathena SJ, Velasquez MT. Beneficial role of dietary phytoestrogens in obesity and diabetes. The American Journal of Clinical Nutrition. 2002;**76**(6):1191-1201

[64] Fukumitsu S et al. Flaxseed lignan attenuates high-fat diet-induced fat accumulation and induces adiponectin expression in mice. British Journal of Nutrition. 2008;**100**(3):669-676

[65] Prasad K, Khan AS, Shoker M. Flaxseed and its components in treatment of hyperlipidemia and cardiovascular disease. International Journal of Angiology. 2020;**29**(04):216-222

[66] Kristensen M, Jensen MG. Dietary fibres in the regulation of appetite and food intake. Importance of viscosity. Appetite. 2011;**56**(1):65-70

[67] Zhu L, Sha L, Li K, Wang Z, Wang T, Li Y, et al. Dietary flaxseed oil rich in omega-3 suppresses severity of type 2 diabetes mellitus via anti-inflammation and modulating gut microbiota in rats. Lipids in health and disease. 2020;**19**(1):1-16

[68] Rahimlou M et al. Effects of flaxseed interventions on circulating inflammatory biomarkers: A systematic review and meta-analysis of randomized controlled trials. Advances in Nutrition. 2019;**10**(6):1108-1119

[69] Ghaseminasab-Parizi M, Nazarinia M-A, Akhlaghi M. The effect of flaxseed with or without anti-inflammatory diet in patients with rheumatoid arthritis, a randomized controlled trial. European Journal of Nutrition. 2022;**61**(3):1377-1389

[70] Ahmadniay motlagh H, Aalipanah E, Mazidi M, Faghih S. Effect of flaxseed consumption on central obesity, serum lipids, and adiponectin level in overweight or obese women: A randomised controlled clinical trial. International Journal of Clinical Practice. 2021;**75**(10):e14592

[71] Lee JC et al. Dietary flaxseed enhances antioxidant defenses and is protective in a mouse model of lung ischemia-reperfusion injury. American Journal of Physiology-Lung Cellular and Molecular Physiology. 2008;**294**(2):L255-L265

[72] Tilg H, Moschen AR. Food, immunity, and the microbiome. Gastroenterology. 2015;**148**(6):1107-1119

[73] Askarpour M et al. Effect of flaxseed supplementation on markers of inflammation and endothelial function: A systematic review and meta-analysis. Cytokine. 2020;**126**:154922

[74] Virani SS et al. Heart disease and stroke statistics—2020 update: A report from the American Heart Association. Circulation. 2020;**141**(9):e139-e596

[75] Burr ML et al. Effects of changes in fat, fish, and fibre intakes on death and myocardial reinfarction: Diet and reinfarction trial (DART). The Lancet. 1989;**334**(8666):757-761

[76] Bang H, Dyerberg J, Nielsen AB. Plasma lipid and lipoprotein pattern in Greenlandic west-coast Eskimos. The Lancet. 1971;**297**(7710):1143-1146

[77] Gonzalez M et al. Defining a relationship between dietary fatty acids and the cytochrome P450 system in a mouse model of fatty liver disease. Physiological Genomics. 2011;**43**(3):121-135

[78] Dupasquier CM et al. Effects of dietary flaxseed on vascular contractile function and atherosclerosis during prolonged hypercholesterolemia in rabbits. American Journal of Physiology-Heart and Circulatory Physiology. 2006;**291**(6):H2987-H2996

[79] Pacheco JT, Daleprame JB, Boaventura GT. Impact of dietary flaxseed (*Linum usitatissimum*) supplementation on biochemical profile in healthy rats. Nutricion hospitalaria. 2011;**26**(4):798-802

[80] Ander BP et al. Differential sensitivities of the NCX1. 1 and NCX1. 3 isoforms of the Na+–Ca2+ exchanger to α-linolenic acid. Cardiovascular Research. 2007;**73**(2):395-403

[81] Ander BP et al. Dietary flaxseed protects against ventricular fibrillation induced by ischemia-reperfusion in normal and hypercholesterolemic rabbits. The Journal of Nutrition. 2004;**134**(12):3250-3256

[82] Caligiuri SP et al. Flaxseed for hypertension: Implications for blood pressure regulation. Current Hypertension Reports. 2014;**16**(12):1-13

[83] Prasad K. Dietary flax seed in prevention of hypercholesterolemic atherosclerosis. Atherosclerosis. 1997;**132**(1):69-76

[84] Toulabi T et al. Effects of flaxseed on blood pressure, body mass index, and total cholesterol in hypertensive patients: A randomized clinical trial. EXPLORE. 2022;**18**(4):438-445

[85] Khalesi S, Irwin C, Schubert M. Flaxseed consumption may reduce blood pressure: a systematic review and meta-analysis of controlled trials. The Journal of nutrition. 2015;**145**(4):758-65

[86] Udenigwe CC et al. Low molecular weight flaxseed protein-derived arginine-containing peptides reduced blood pressure of spontaneously hypertensive rats faster than amino acid form of arginine and native flaxseed protein. Food Chemistry. 2012;**132**(1):468-475

[87] Ursoniu S et al. Effects of flaxseed supplements on blood pressure: A systematic review and meta-analysis of controlled clinical trial. Clinical Nutrition. 2016;**35**(3):615-625

[88] Khalesi S, Irwin C, Schubert M. Flaxseed consumption may reduce blood pressure: A systematic review and meta-analysis of controlled trials. The Journal of Nutrition. 2015;**145**(4):758-765

[89] Hadi A et al. Effect of flaxseed supplementation on lipid profile: An updated systematic review and dose-response meta-analysis of sixty-two randomized controlled trials. Pharmacological Research. 2020;**152**:104622

[90] Bassett CM, Rodriguez-Leyva D, Pierce GN. Experimental and clinical research findings on the cardiovascular benefits of consuming flaxseed. Applied physiology, nutrition, and metabolism. 2009;**34**(5):965-974

Chapter 7

Lignin as Feedstock for Nanoparticles Production

Ana Lourenço and Jorge Gominho

Abstract

Lignin is an interesting natural polymer with characteristics that contribute for the development and growth of plants. Lignin presents high variability associated with the diversity of plants, which presents great challenges for its recovery after delignification (*technical lignin*), because lignin is prone to irreversible degradation, producing recalcitrant condensed structures that are difficult to disassemble afterward. Although researchers have made efforts to obtain lignin in high yields and with good characteristics for specific uses, this is not an easy task. The mind-set has changed and new biorefinery concepts are emerging, where lignin is the primary goal to achieve, and the so-called *lignin-first* approach has arisen. Lignin can be obtained firstly to prevent structural degradations, enabling an efficient and highly selectivity of the lignin monomers. Therefore, this concept places lignin and its valorization at the head of the biorefinery. However, lignin valorization is still a challenge, and to overcome this, lignin nanoparticles (LNPs) production presents a good way to achieve this goal. This chapter presents a resume of the several techniques to attain lignin, how to produce LNPs, and their possible applications (from pharmaceutical to the automobile and polymer industries).

Keywords: biomass composition, lignin isolation, *lignin-first* biorefining, lignin applications, nanoparticles production

1. Introduction

Lignin has been studied by different researchers over the years; the first output was published in 1908, and to date, a total of 74,191 documents mentioning "lignin" have been published, according to the Web of Science database (WOS). **Figure 1** presents a treemap with the published documents distributed by the WOS categories (from 1908 to 2023): Biotechnology Applied Microbiology (9440), Energy fuels (9362), Materials Science Paper Wood (9256), Chemistry Multidisciplinary (8935), Chemical Engineering (8564), to mention just a few.

A search for the word "lignin-first" revealed a smaller number of documents, totaling 127. But what was quite impressive was the date of the first document, in 1997 (**Figure 2**). The main articles were published in the last 6 years, reaching a total of 122. This fact reveals the increasing importance of the "lignin-first" approach within the research community.

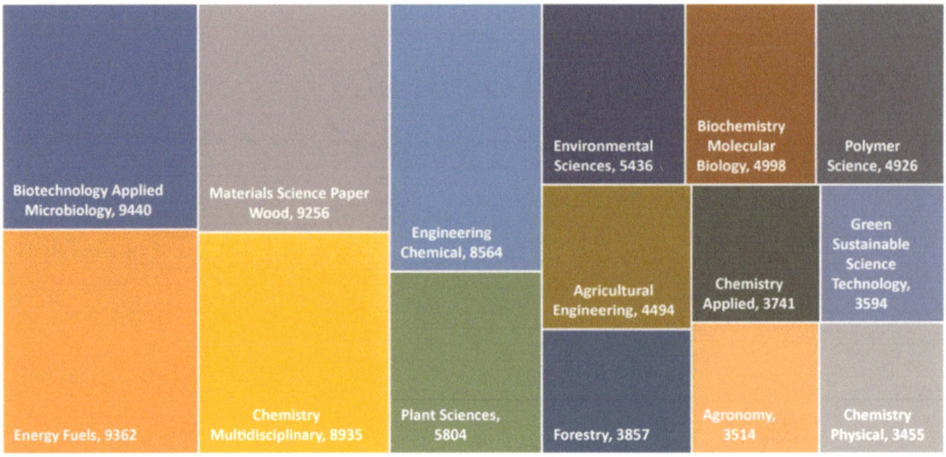

Figure 1.
TreeMap chart adapted from WOS searching results using the word "lignin" (www.webofscience.com; accessed on September 30, 2022).

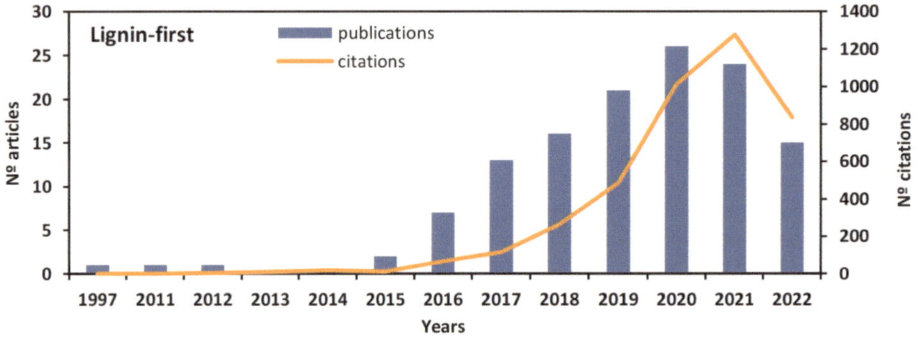

Figure 2.
The number of publications that mention the word "lignin-first" and their citations over the years. Adapted from WOS (www.webofscience.com; accessed on September 30, 2022).

The topic "lignin nanoparticles" is also an increasing one (**Figure 3**), and the search provided a total of 373 documents during the period from 2008 to 2022 (September). The most highly cited papers number 10, with 332 articles and 32 review papers recorded. The distribution between the WOS categories was as follows: Chemistry Multidisciplinary (123), Polymer Science (81), Materials Science Multidisciplinary (59), and Biochemistry Molecular Biology (51). Although the year 2022 has not ended, it is possible to see the growing tendency for the number of publications and citations of this topic (**Figure 3**).

According to these results, it is clear that lignin is still an exciting research topic. To reinforce this, at the moment, a European Cooperation in Science & Technology is being developed within this theme, called "*LignoCost - Pan-European Network on the Sustainable Valorisation of Lignin*" (COST Action CA17128, https://lignocost.eu/), to establish a network between different researchers having a focus on lignin valorization and its sustainable industrial application. One of the Working Groups is devoted to making a Wikilignin database where relevant works are being compiled.

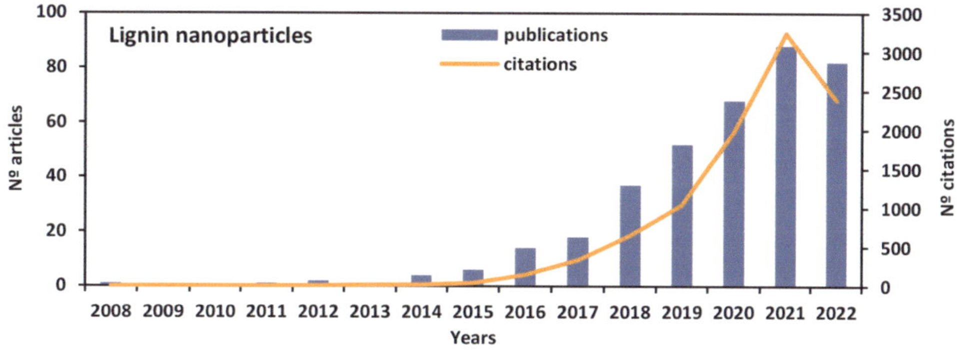

Figure 3.
The number of publications that mention "lignin nanoparticles" and their citations over time. Adapted from WOS. (www.webofscience.com; accessed on September 30, 2022).

2. Lignin structure and composition in biomass

2.1 What is lignin?

Lignin is a complex polymer present in vascular plants, particularly in woody tissues. It is the second most abundant natural polymer, representing 15–30% dry weight of lignocellulosic materials (e.g., wood and bark). Its natural production on earth ranges between 5 and 36 $\times$ 10^8 tons annually [1]. Lignin is mainly deposited in the secondary cell walls of the supporting and water-conducting tissues, but it is also present in the middle lamella between cells and cell corners. Besides xylem cells, lignification can also occur in sclereids cells (phloem cells), roots, and seeds [2].

2.2 What is lignin made of?

Lignin is a heterogeneous aromatic polymer constituted by three phenylpropane precursors: *p*-coumaryl alcohol, coniferyl alcohol, and sinapyl alcohol that present a hydroxyl group linked to C4 and substitutions with one or two methoxyl groups in C3 and C5. If the aromatic ring is unmethoxylated, then it is called *p*-hydroxyphenyl (H), and if it has one methoxyl group in C3, it is called guaiacyl, or with two methoxyl groups linked with C3 and C5, it is a syringyl unit (S, **Figure 4**) [3]. Therefore, lignin is a highly complex phenolic polymer whose content and composition vary depending on many factors, for example, species, specimen age, plant part, environmental growth conditions, and harvesting time [2–7]. In hardwoods, lignin typically represents between 20 and 25% of the wood, with a monomeric composition mainly of syringyl and guaiacyl but few *p*-hydroxyphenyl units. In comparison, softwoods are richer in lignin (25–35%), where guaiacyl units prevail, fewer *p*-hydroxyphenyl units are found, and almost no syringyl units are present [3]. In herbaceous species, the lignin content ranges from 15 to 24% [8] and has a monomeric composition where no specific lignin units predominate [3]. It was found that other monomers are also included in the lignin polymer, such as tricin flavones, hydroxycinnamaldehydes, hydroxystilbenes, and xenobiotics [9–15]. The monomers are linked by ether bonds and carbon-to-carbon bonds in different proportions. The ether bonds prevail (e.g., β—O—4′ and α—O—4′), representing between 35 and 60% in softwoods and from 50 to 80% in hardwoods [16]. Generally, C—O linkages have a bond-dissociation

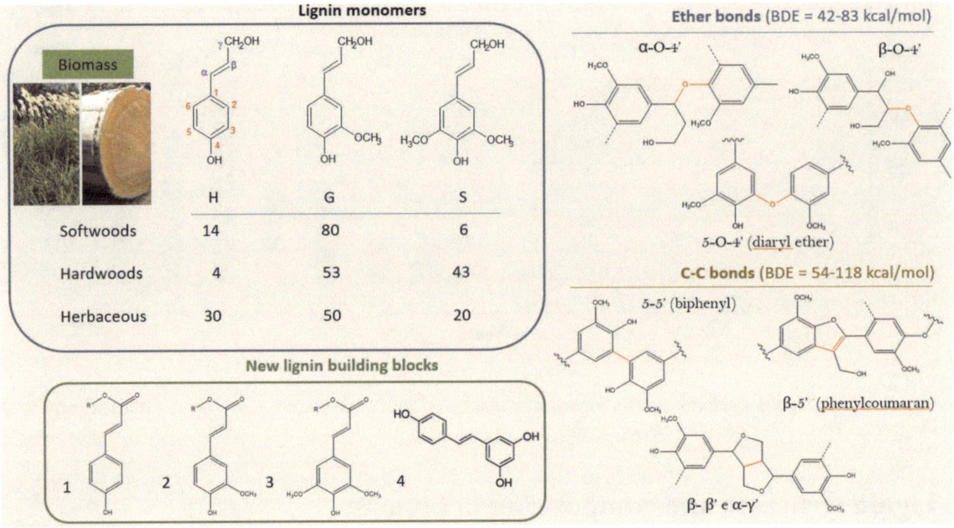

Figure 4.
Lignin monomeric composition distributed by softwood, hardwood, and herbaceous. The new lignin building block: 1—coumaric acid, 2—ferulic acid, 3—sinapic acid, if R=H; 4—resveratrol (hydroxstilbene).

enthalpy (BDE) of 42–83 kcal/mol (they are easier to break); while C—C linkages (e.g. β—β′, β—5′) have BDE values between 54 and 118 kcal/mol, so they are harder to break down, presenting more considerable challenges for lignin deconstruction [17]. Slightly different values were compiled by Liu et al. [18] and are presented in **Table 1**.

Component	Type	Total amount (%)			Bond dissociation enthalpy (BDE) (kcal/mol)
		Softwood	Hardwood	Grass	
Monolignol	H	<5	0–8	5–33	—
	G	>95	25–50	33–80	—
	S	0	46–75	20–54	—
Linkages					
C—O—C	β-O-4′	43–50	50–65	74–84	56.54–72.30
	α-O-4′ *	5–7	<1	n.d.	48.45–57.28
	4-O-5′	4	6–7	n.d.	77.74–82.54
C—C	5-5	5–7	<1	n.d.	114.9–118.4
	β-β	2–6	3–12	1–7	—
	β-5	9–12	3–11	5–11	125.2–127.6
	β-1	1–9	1–7	n.d.	64.7–165.8
Others		16	7–8	n.d.	—

**Only present in the dibenzodioxocin moieties (β—O—4′ + α—O—4′ + 5-5).*

Table 1.
General characteristics of lignin from softwood, hardwood, and herbaceous species. Adapted from Liu et al. [18].

2.3 What is the role of lignin in plants?

Lignin deposition in plant cells: (i) provides rigidity and mechanical strength, allowing the plants to stand; (ii) gives hydrophobicity to the cell walls; (iii) promotes the transportation of water and minerals between cells, (iv) hinders the degradation of polysaccharides; and (v) is a barrier against pathogens and pests, playing an essential role in plant development [19–21].

Lignin has another active role in the plant response to several abiotic and biotic stresses, by reinforcing the cell walls in the neighborhood of infected or wounded tissue (the so-called *defense lignin*, which contrasts with the *developmental lignin* presented before) [22–24]. Plants have suffered different types of abiotic (*e.g.*, drought, mineral deficiencies, and low or high temperatures) and biotic stresses (*e.g.*, insects, bacteria, or fungi attacks) over the years. Lignin is vital for plant growth and its environmental adaptability, that is, the plant can produce different amounts and types of lignins (possessing other physical properties) depending on the growth conditions. Moura and coworkers [25] reported a review explaining the influence of different stresses, such as the increase of cold, drought, or light upon the plants; for example, maize under water deficit reduced the production of ferulic acid but increased *p*-coumaric and caffeic acids in the xylem regions. However, the roots increased their lignin content, stiffening the cell walls and reducing their expansion [25]. More recently, a review was presented focused on the biosynthesis, content, and accumulation of lignin as a plant response to abiotic stresses; during drought, *E. urograndis* increased the lignin content in its roots and leaves and reduced the S/G ratio [26]. In *P. trichocarpa,* the lignin content did not change in young shoots or mature stems, but the S/G ratio decreased significantly in the young shoots. Plants under drought stress may increase the lignin content to reduce water penetration and transpiration from the cell wall, which helps to maintain cell osmotic balance [21].

Plants under a rise in light induced a lignin accumulation, as a way to adapt to the environment [26]. Also, the lignin composition in angiosperm and gymnosperms species under stress revealed a higher amount of condensed linkages (C—C bonds) and more H-units as a response to the environmental conditions, including high nitrogen fertilization, mechanical injuries, or ozone [2, 23, 24].

Therefore, researchers have tried to understand plant adaptability, but it is difficult to generalize the behavior of plants regarding particular stress because each species behaves differently. Also, not only the lignin content or its composition is essential to understand plant behavior, since the production of lignin comprises a complex genetic network with the involvement of multiple enzymes that have different responses to the abiotic and biotic factors [24]. This will not be developed here, but readers are encouraged to consult many of the works on this topic [27–30] and a review about lignin biosynthesis [31].

The diversity of lignins (content and composition) and their recalcitrant behavior to degradation is a challenge for researchers who wish to learn more about the lignin structure and biosynthesis, but mainly for those who work in lignin valorization. Thereby, we discuss below some procedures to isolate the lignin.

3. Lignin isolation procedures and *lignin-first* concept

Most often, lignin valorization starts with its isolation from the lignocellulosic material (biomass) with two perspectives: (i) to gain insights into lignin's chemical and structural characterization, which involves methods such as Björkman, dioxane, and cellulolytic enzyme lignin [32–34], where the obtained lignin presents a structure close to the lignin in the cell wall matrix (*native lignin*); or (ii) aiming to obtain lignin in a high yield, in which case the biomass is delignified (pulp process), obtaining a solid fraction rich in cellulose and a liquid one mainly constituted by lignin and hemicelluloses. Afterward, the lignin is isolated by precipitation from the spent liquor (Kraft, organosolv, or lignosulfonate) in yields dependent on the precipitation agent, pH, etc. In these cases, the pulping process leads to an extensive chemical change in the lignin, such as the cleavage of β—O—4′ linkages and the formation of recalcitrant C—C bonds, which results in difficult conditions for lignin depolymerization. This lignin presents different characteristics (related to the isolation process) from those of *native lignin* and is called *technical lignin* [35].

There are several pulping processes from which different technical lignins can be obtained [8, 36–38]. The conditions applied depend on the biomass chemical characteristics, particularly the lignin content and composition, but these issues will not be discussed here. Instead, a brief description of the processes is presented in the following lines.

Soda pulping started in 1853 using non-woody biomass such as straw or flax, but now the process has been applied to soft and hardwoods. The delignification is done with sodium hydroxide (NaOH) liquor under temperatures around 160°C. Lignin depolymerization is promoted, starting with the cleavage of β—O—4′ linkages, which allows the ionization of phenolic groups, but condensation reactions also occur that are negative for the further lignin valorization. Besides, the lignin-carbohydrate complexes (LCCs) are also broken [18].

Kraft pulping was developed later than soda pulping (around 1890) and has become the most common process involving the delignification of either softwood or hardwood, but also non-woody species, with a white liquor containing sodium hydroxide (NaOH) and sodium hydrosulfide (Na_2S) at temperatures around 170°C [39, 40]. Delignification occurs by the hydroxyl groups (OH^-) but also by hydrogen sulfide (HS^-), which, due to its nucleophilic character, enhances delignification. Lignin has been extracted industrially from the black liquor by *LignoBoost* or *LignoForce* technology [41]. It is the most commonly used market process to produce technical lignins [18]. The isolated lignin presents sulfur as thiol groups (which make lignin valorization difficult), is highly condensed, and contains a low amount of ether bonds [8].

Sulfite pulping started in 1930. Biomass is pulped under the attack of sulfite or bisulfite salt of sodium, ammonium, or magnesium at temperatures ranging from 140 to 170°C. The sulfonic groups are incorporated in the aliphatic chain of the lignin monomers, and water-soluble lignosulfonate salt is formed. The *technical lignin* produced is called lignosulfonate, representing 90% of the commercialized lignin [34]. Still, since Kraft pulping is more efficient, the sulfite process has decreased (<5%) [8]. The lignosulfonate: (i) can be isolated by ultrafiltration, extraction, or precipitation; attained as a salt (Na^+, NH_3^+, Mg^+, Ca_2^+); (ii) presents a large amount of sulfur (4–8% wt compared with Kraft lignin), and has a high number of degraded oligomers and a low amount of β—O—4′ [8].

Organosolv pulping appeared in the late twentieth century as a promising process to selectively attain lignin using, for example, methanol or ethanol. The lignin extraction is done by solvolysis with or without the presence of acids (*Brønsted*, e.g., H_3PO_4,

H_2SO_4, AcOH, CF_3SO_3H; or *Lewis acids* such as $FeCl_2$, $ZnCl_2$). If the acid is not used, the pH will decrease from 7 to 4 due to the hemicelluloses deacetylation; on the contrary, if the acid is present, the pH can decrease to between 4 and 2; thus, the solvolysis of the biomass is accelerated, and both hemicelluloses and lignin are removed [17].

Ionic liquids (IL) have been used for lignin extraction due to their characteristics (e.g., green and recyclable). Overejo-Pérez and coworkers [42] used the protic ionic liquid 1-methylimidazolium chloride to isolate lignin from *E. globulus* under different conditions to study the treatment severity. More severe conditions promoted: (i) the cleavage of ether linkages, (ii) lignin fractions with higher molecular weight (MW) and more stable due to condensation reactions, and (iii) preferential degradation of S-units, contrasting to mild conditions where the degradation is of G-units.

Deep eutectic solvents (DESs) are promising alternative solvents for IL because they can also be recycled, are biodegradable, easy to produce, and reuse several times without losing strength, depending on the chosen system. Lignin from corncob residues was extracted by DES and used to make LNP by self-assembly [43].

Therefore, there is a great diversity of the *technical lignins* obtained due to the intrinsic variability of the biomasses used and the different extraction processes. As a result, the heterogeneity of the *technical lignins* is a negative point for their valorization. As an example, **Table 2** presents the characteristics of Kraft lignin and lignosulfonates.

Consequently, upgrading solutions must be found so that the concepts of biorefinery, the circular economy, and the zero-waste philosophy can be fully applied. Having in mind these aspects, researchers have treated biomass under mild conditions to attain more uniform lignin products in the so-called *lignin-first* approach (also known as reductive catalytic fractionation (RCF) or catalytic upstream biorefining (CUB)). Under this new biorefinery concept, the lignin is firstly removed from the biomass (instead of the carbohydrates fraction), and the cellulose and hemicellulose fractions are almost intact [44, 45]. Abu-Omar and coworkers [7] present the concept: "*The lignin-first biorefining is not a synonym for lignin valorization, but rather an integral approach that derives value from both lignin and polysaccharides, towards an atom-efficient and more sustainable utilization of lignocellulosic biomass*". This approach is more selective for lignin; hence, it prevents undesirable and irreversible condensation reactions, eliminates the need for purification steps, and reduces production costs [45]. The traditional methods promote the cleavage of the β—O—4′ linkages and the formation of C—C bonds,

Properties	Kraft lignin	Lignosulfonates
Functional groups	Many phenolic OH, carboxyl, catechol groups	Fewer phenolic OH, carboxyl, catechol groups
Organic sulfur (%)	1–1.5	4–8
Sulfonated groups (m_{eq}/g)	0	1.25–2.5
Solubility	Alkali, acetone, DMF, methyl cellosolve™	Water

Table 2.
Properties of Kraft lignin and lignosulfonates. Data from Bozell et al. [34].

producing a condensed lignin that is harder to depolymerize in successive steps (**Figure 4**, fragments E and F). To overcome these negative aspects, the *lignin-first* approach involves the application of mild conditions to stabilize the β—O—4′ linkages and the low lignin MW products (e.g., monomers, dimers, and short oligomers) formed during the biomass fractionation, preventing lignin condensation [46].

In summary, the *lignin-first* approach has three steps: (i) the lignin is extracted from the biomass by an organic solvent through solvolysis or acid-catalyzed reactions (similarly to organosolv pretreatment); (ii) the intermediates formed are stabilized to prevent condensation, and (iii) the lignin can be further depolymerized during the stabilization stage [7]. The stabilization approaches deliver the target chemical molecules without requiring further chemical modifications [47].

Some biomass fractionation methods applied to obtain uncondensed lignin include (i) ammonium-based fractionation—which can solubilize the lignin at room temperature under a pressure of 7–10 bars [48]; (ii) mild organosolv—using organic solvents (e.g., ethanol, methanol, acetic and formic acids) that are recycled during the recovery of the lignin [49]; (iii) γ-valerolactone-assisted hydrolysis—using γ-valerolactone (GVL, a green solvent that can be produced from glucose) and water (cosolvent), for example, pine wood fractionated by 80% γ-valerolactone at 140–180°C, yields lignin of 33% [50]; or (iv) ionic-liquid-assisted fractionation [51].

Other methods are more focused on stabilizing the lignin monomers; the most studied one is catalytic hydrogenolysis, which combines lignin depolymerization with solvolytic extraction under the reductive stabilization of the intermediates. The hydrogenolysis cleaves the C—O bond at the β—O—4′ moieties, generating lignin monomers; the reaction requires a solvent that will be responsible for the lignin

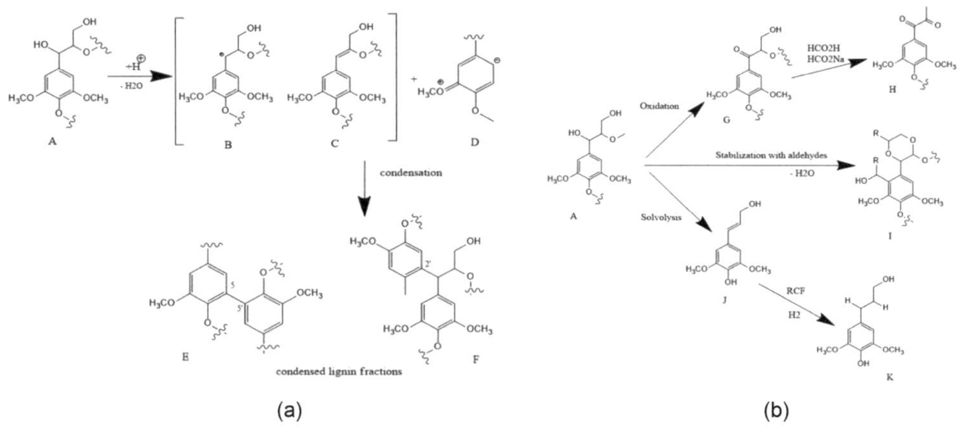

Figure 5.
Lignin depolymerization by acid hydrolysis: (a) hydrolysis and dehydration of the β—O—4′ lignin fragment A, leads to fragments B (benzylic carbocation intermediate), C (alkene products from the acid hydrolysis). All these fragments can react, and condensation reactions occur, leading to products E (includes an α-2′ linkage) and F (includes a 5-5′ linkage). (b) The β—O—4′ lignin fragment can be involved in the following reactions: (i) oxidation leading to fragment G (oxidized lignin fragment) that can form fragment H (oxidized monomer), (ii) solvolysis leads to products that can be transformed by reductive catalytic fractionation into fragment K (a reduced monomer), and (iii) stabilization with aldehydes forms the acetal-stabilized lignin fragment (I). Adapt from Questello-Santiago et al. [17].

depolymerization (e.g., methanol or formic acid that will be the H_2 donor), and a metal-containing catalyst such as Pt, Ru, Pd, Ni, or Rh, to prevent lignin condensation [17].

Shuai et al. [52] and other researchers proposed another *lignin-first* strategy to preserve the C_{aryl}-O ether bonds, particularly the β—O—4′ moieties, to prevent the formation of a Benzylic carbocation intermediate (**Figure 5a**, fragment B) due to its condensation ability. During lignin extraction, the β—O—4′ linkage is stabilized with an aldehyde (e.g., formaldehyde—CH_2O or acetaldehyde—MeCHO) to form the stable acetals, that is, cyclic 1,3-dioxanes (**Figure 5b**, fragment I) that entrap the diol, preventing its dehydration and degradation. This method permits the dissociation between the biomass fractionation from the subsequent depolymerization step, enabling: (i) a broader range of depolymerization methods; and (ii) the optimization of the fractionation and depolymerization steps since they are independent. The advantage is that this produces insignificant condensation reactions, a high yield of monomers, and high selectivity. However, the method presents some limitations for large-scale implementation: (i) it is challenging to find the right balance between the removal of lignin from the biomass and the preservation of the carbohydrates in the pulp; (ii) the separation of the catalyst from the pulp. Both points are seen as a downside of this method. Questello et al. [17] attained a lignin monomeric yield of 42–50%, but it was structurally more complex (**Figure 5b**, fragment G).

Other critical reviews were published in 2020 focused on developments in the *lignin-first* approach [18, 35, 44, 53, 54], in 2021 [7, 38] and in 2022 [45, 55]. One of them emphasizes the fundamental catalytic reactions of the extraction and depolymerization of lignin and posterior stabilization of the phenolic units; the authors also present a brief overview of the possible modifications of the lignin-derived phenols and monolignols, focused on added-value chemicals, polymers, and other developments [53]. Gigli and Crestine [35] present different methodologies for

Fractionation methods	Method	Possible procedure
pH	Alkali soluble lignin: change pH > 12 to 1	1. Sequential lignin precipitation by H_2SO_4 or HCl (room temperature) 2. Precipitation with acid and separation of the insoluble part 3. Separation by pressure at high temperature
Solvent/water extraction	Partial solubility in different solutions (organic solvents/water)	1. Ethanol/water 2. Acetone/water 3. Tetrahydrofuran (THF)/water 4. Acetic acid (ACOH)/water (room temperature) 5. AcOH/water (high temperature & pressure) 6. γ-valerolactone (GVL)/water
Solvent extraction	Partial solubility in organic solvents	Alcohols, chloroform, ether, dimethyl sulfoxide (DMSO), dichloromethane (DCM), THF, hexane, ketones
Ultrafiltration/membrane	Physical filtration	1. Ceramic filters 2. Polymeric filters

Table 3.
Methods for the fractionation of lignin. Adapted from Sadeghifar and Ragasukas [56].

lignin fractionation by (i) solvents (most common)—using solvent mixtures or under a sequence of solvents; (ii) membranes—based on ultra- and nanofiltration using ceramic or polymeric membranes; (iii) precipitation of the re-dissolved lignin in a binary solvent system; and (iv) pH-mediated gradient precipitation (**Table 3**). The lignin-membrane-based approach makes the separation into cuts of lignin with a defined molecular weight (MW) and narrow polydispersity, while the goal of the solvent-based approach is to obtain fractions with distinctive physicochemical properties. However, both parameters (MW and structure/functionalities) appear to be closely interdependent, since low MW fractions: (i) are typically soluble in polar solvents; (ii) possess high phenolic content and condensed structures; and (iii) have increased antioxidant activity. On the other hand, the higher MW fractions present fewer modified lignins but have a lower amount of aromatic hydroxyl groups and condensing units. The authors also highlight the applications of fractionated lignin, keeping in mind their characteristics. For example, lignin with high MW attained from *P. radiata* Kraft black liquor was successfully used to produce phenol-formaldehyde resins after methylolation and demethylation (to increase the lignin reactivity) [57].

The main lignin applications studied so far are: (i) polymers production such as composites (17%), thermosets (8%) and thermoplastics (4%); (ii) properties improvement involving antioxidant activity (15%), standards generation (6%), antimicrobial activity (4%) and color reduction (2%); (iii) micro and nano-structures such as fibers (12%) and particles (6%); (iv) others (14%) and finally (v) structure modification (12%) [35].

A state of the art on lignin valorization strategy was presented by Liu et al. [18], who discuss the production of *technical lignins* (giving a resumé of the fractionation methods such as Kraft or organosolv delignification), describe conventional methods for lignin catalytic depolymerization (e.g., pyrolysis, solvolysis, catalytic oxidative and reductive depolymerizations), plus the emerging strategy of the *lignin-first* approach, and the methods to improve the yield of the lignin phenolic products. Karlsson et al. [54] present a study where green solvents are used to obtain high yields of the fraction's hemicelluloses lignin and cellulose. The biomass was treated by supercritical water extraction to remove the hemicelluloses (liquid fraction, characterized by a partial hydrolysate of the glycosidic bonds, but preserving the native structure of the oligomers). Then the solid fraction was treated with aqueous ethanol supercritical extraction catalyzed with H_2SO_4 at 160°C to remove the lignin (liquid) and the solid rich in cellulose (with crystallinity preserved after both treatments). Overall, the proposed method combines active stabilization with physical lignin protection, minimizing the lignin condensation reactions. The potential uses of each fraction were discussed: hemicelluloses and cellulose could be used for ethanol production or hydrolyzed to monomeric sugars for chemical applications; the lignin could be used directly for polymers production or be further depolymerized to the monomeric platform [54]. Abu-Omar and coworkers [7] discuss the importance of biomass diversity and the analytical methods used for its characterization (e.g., analytical pyrolysis, NMR), but also present the big picture of the *lignin-first* approach, from feedstock to reactor design, the importance of the catalyst, the mass balance and yields, and finish with an outlook on the development of *lignin-first* biorefinery [7]. Korányi et al. [44] presented an extensive list of the works done under the *lignin-first* approach from 2018 to 2020, focusing also on the studies on wood digestion/conversion from 1940 to 2014, mentioning the different feedstock, conditions, and products obtained. As mentioned before, the

Lignin chemistry	Possible procedure
Base-catalyzed reactions	1. Alkaline conditions enable: (i) cleavage of lignin-carbohydrate (LCC) bonds; (ii) cleavage of ether bonds ⇒ lignin fragmentation; 2. Distinctive reactivity of β—O—4′ bonds, cleaved on: (i) phenolic units (free phenolic OH groups) ⇒ easily cleaving to depolymerize the lignin (the unit is transformed into a quinone methide (if the Cα has a suitable leaving group like —OH or —OR). Quinone methide can undergo nucleophilic attack by HS- (Kraft pulping), with the prevailing pathway of β—O—4′ cleavage and formation of an episulfide intermediates (can undergo some reactions forming coniferyl alcohol and others ⇒ prone to degrade and repolymerize); (ii) non-phenolic units (etherified phenolic OH groups) ⇒ β—O—4′ bonds are slowly cleaved and involve the formation of an epoxide intermediate; 3. Third pathway involves the elimination of the $C\gamma H_2$-OH group (retro-aldol reaction) ⇒ formation of an enol ether structure (alkali stable) and formaldehyde (C_2HO, during soda pulping). In this case, ether bonds are cleaved but with less efficiency. C_2HO induces repolymerization via formaldehyde-phenol type of condensation. **Disadvantage**: possible lignin degradation and repolymerization.

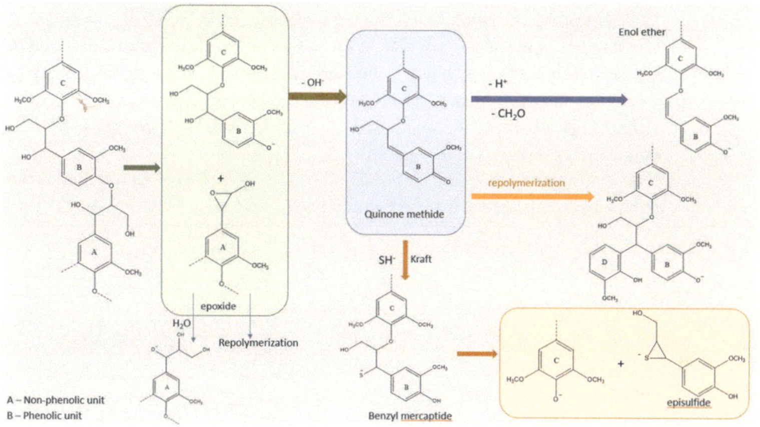

Table 4.
Possible chemical reactions during lignin fractionation by base solvents. Adapted from Schutyser et al. [8].

characteristics of the technical lignins depend on the methods used to isolate or depolymerize the lignin. Therefore, **Tables 4–6** present a resumé of the lignin chemistry during the different processes.

Lignin chemistry	Possible procedure
Acid-catalyzed reactions	1. Promote hydrolysis of ether bonds in cellulose and hemicelluloses, but also the acidolysis of lignin ⇒ depolymerization and repolymerization reactions; 2. Cleavage of ether bonds (β—O—4′) ⇒ removal of OH in Cα, formation of a benzylic carbenium ion that can be transformed into two enol-structures with or without the cleavage of Cβ—Cγ bonds and concurrent formation of formaldehyde; 3. Using H_2SO_4 promotes enol-structure without the bond Cβ—Cγ; using HCl or HBr ⇒ enol-structures formation; 4. The hydrolysis of the enol-structures leads to C2-aldehyde-substitutes or C3-ketone-substituted phenolics (Hibbert's ketones). **Disadvantage:** Hibbert's ketones structures and carbenium ions can produce condensed structures ⇒ lignin condensation.

Lignin chemistry	Possible procedure

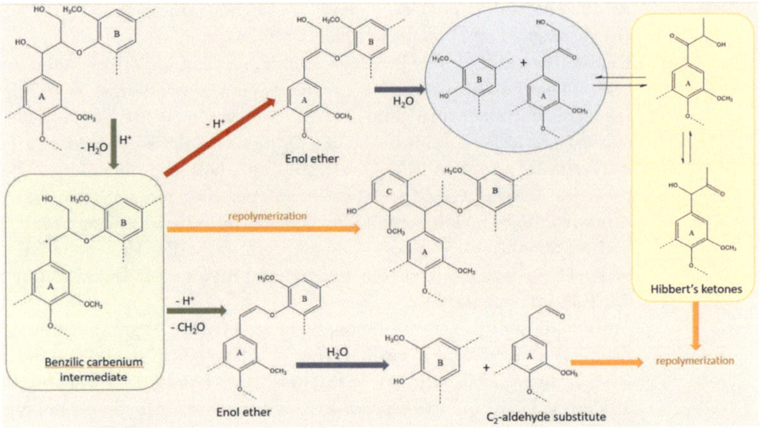

Reductive reactions	1. Redox catalyst & combination with H_2 or H-donors ⇒ targets: cleavage of ether bonds (β-O-4′ and α-O-4′) and side chain OH groups; 2. Other pathways result in: (i) hydrogenolysis of ether bonds; (ii) removal of benzylic OH–groups in Cα and possibly in Cγ ⇒ formation of substituted methoxyphenols and small fragments (e.g. phenol). Secondary reactions can occur, such as hydrogenation/hydrogenolysis ⇒ production of cyclohexanols and cycloalkanes. **Advantage**: prevents the formation of reactive functional groups ⇒ avoiding condensations or repolymerization. **Disadvantage**: inability to cleave C—C bonds ⇒ degree of depolymerization is linked to the amount of ether bonds that can be cleaved.

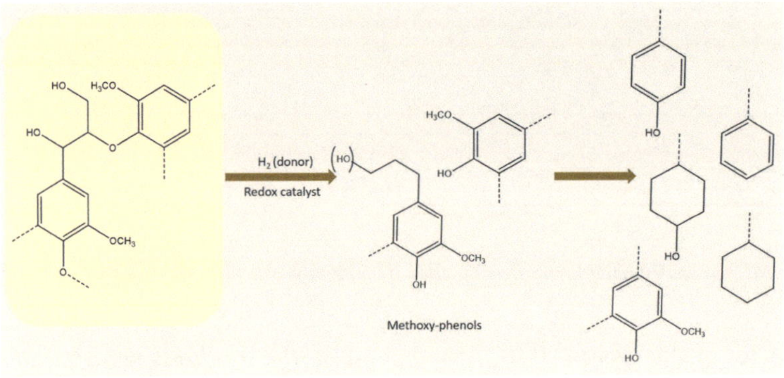

Table 5.
Possible chemical reactions during lignin fractionation by acid solvents and reductive reactions. Adapted from Schutyser et al. [8].

Despite all the work developed so far, there are still some issues to be overcome for the large-scale implementation of the *lignin-first* approach, such as that (i) the lignin fractions are correlated with the native lignin, that is, depend on the biomass used, so obtaining fractions with high reproducibility is an issue; (ii) the fractioning process is usually done in a batch, but a continuous process is more favorable for industrialization and to use the black liquor directly from the pulping industry [35].

Lignin chemistry	Possible procedure
Oxidative lignin chemistry	1. Pathway: (i) fragmentation of the side chain (cleavage of Cα—Cβ of Cα—C4⇒ unstable structures are formed under alkaline conditions, thus can be converted to aliphatic carboxylic acids and the aromaticity is retained; (ii) the aromatic ring is cleaved ⇒ yields aliphatic carboxylic acids. **Advantage**: cleavage of C—C bonds (instead of ether bonds, as in the other reactions mentioned). **Disadvantage**: the reaction follows a radical mechanism, so radical coupling of lignin fragments takes place ⇒ producing biphenyl structures (condensed structures).

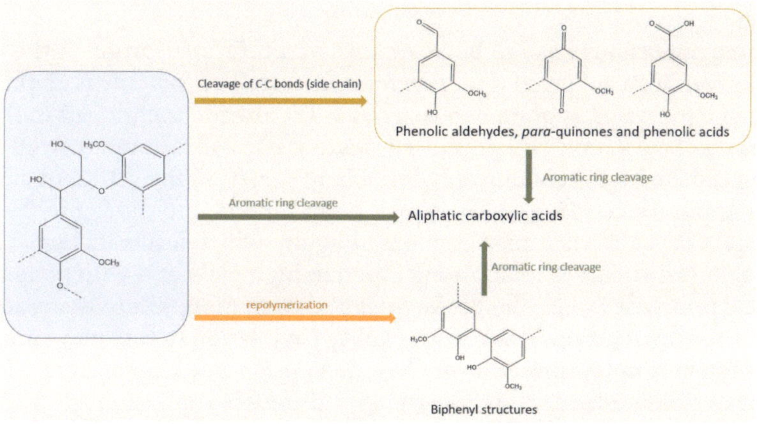

Table 6.
Possible chemical reactions during lignin fractionation by oxidative reactions. Adapted from Schutyser et al. [8].

4. Lignin applications: focus on nanoparticles production

Lignin is an exciting polymer due to its nontoxicity and bio-based renewable nature, making it a sustainable feedstock that does not compete with food chain production [58], and recent technologies to recover lignin and its conversion into valuable compounds have been reviewed [59]. However, lignin applications on a large scale present some limitations due to lignin's poor solubility in water and complex macromolecule structure [35].

However, many efforts have been made; for example, the *Borregaard company* commercializes tailor-made lignin that can be used in wide-ranging industries, as a binding, dispersing agent, but also as an emulsion stabilizer (https://www.borregaard.com/product-areas/lignin-biopolymers/). The *Bloom company* uses aldehyde-assisted fractionation to obtain a stabilized lignin, delivering lignin as a biopolymer with known characteristics that can be used for the production of nutraceuticals, food additives, or even cosmetics (https://www.bloombiorenewables.com/).

Despite these good examples of lignin valorization, in both cases, the lignin is tailor-made, but what about the other lignins produced on a laboratory scale but with high diversity? They have not yet reached the market, in particular because the lignin is obtained in low yields, with structural diversity, and some processes also use dangerous solvents. So, the target of many researchers has been to modify the lignin, or to develop lignin nanoparticles (LNPs), that is, to obtain lignin in the form of aqueous nanoparticles with uniform size and shape, showing better characteristics such as a higher surface area and stability in an aqueous medium, that are non-toxic (in reasonable concentrations), biodegradable (either by bacteria but also fungi), and present bioactive properties, which can replace the commercial lignin [60].

Furthermore, LNP has functional groups that can be chemically modified, increasing their applicability, and can be applied in a wide range of applications [61–92]: (i) encapsulation with biocides or anticancer drugs [62–68], (ii) incorporated in coatings for wood UV protection [69], or in polymeric matrices to enhance mechanical properties (e.g., strength and toughness), barrier properties and thermal stability [70, 71], and (iii) in the agriculture field [72]. **Table 7** presents some LNP applications based on the starting technical lignin.

Still, some limitations have to be overcome, and LNP application at the industrial level has not yet been achieved due to diverse factors. The *first* one is about lignin's variability in nature that is quite heterogeneous (as already mentioned in this chapter); therefore, its isolation will produce a mixture of products, apart from being a challenging task and less economically feasible to purify all the different compounds for further applications [45, 71]. The *second* point is related to the lignin fractionation methodologies that yield complex condensed lignin, which limits its high-value applicability [71, 94], difficulties in attaining lignin in high yield and with good chemical and physical properties (e.g., molecular weight distribution, solubility, reactivity and number of functional groups), and with quality (in relation to the inter-units linkages, condensed lignin is not attractive) [43, 95]. The *third* point is associated with LNP production and their adequate stabilization for different applications [96]. This last point will be developed in the following paragraphs.

Technical lignin nanoparticles	Applications	Ref
Soda lignin	Packaging, agriculture	[73]
	Wastewater treatment	[74]
	Biomedicine	[75, 76]
	Coatings	[77]
	Cosmetics	[78]
Alkali lignin	Cosmetics	[79]
	Wastewater treatment	[80, 81]
	Biomedicine	[82, 83]
	Coatings	[84]
Kraft lignin	Emulsion stabilization & biocatalyst	[85, 86]
	Biomedicine	[62, 63, 87, 88]
	Wastewater treatment	[89]
	Dispersants	[90]
Organosolv	Nanocomposites	[91, 92]
	Cosmetics	[78]
Hydrolytic lignin	Paint and coating	[69]
	Enzyme immobilization	[93]

Table 7.
Possible applications of the different nanoparticles produced from technical lignins. Adapted from Ekielski and Mishra [38].

4.1 Techniques used for the synthesis of LNP

Kumar et al. [58] present an interesting resumé of the several strategies used to produce lignin micro and nanoparticles. Diverse techniques can be used and include: shifting methods (by pH or solvents) [97], a template-based synthesis technique (or polymerization), aerosol process, electrospinning, supercritical fluid processes, solvent antisolvent precipitation [98], or acoustic cavitation. A resumé of techniques to attain LNP is presented below.

Self-assembly—is the most common method to produce LNP; it starts by dissolving the lignin in an organic solvent and by adding the antisolvent (usually water) or, during dialysis, the particles start to form in a self-assembling process.

Antisolvent precipitation—first the lignin is dissolved in a solvent, such as: acetone solution (9:1, v/v) [99]; tetrahydrofuran (THF) [62]; dimethylsulfoxide (DMSO) [92]; dimethylformamide (DMF) [100]; and aqueous sodium *p*-toluenesulfonate (APS) [101]; then this solution is added slowly to an antisolvent (usually water, where the lignin is insoluble) to form the LNP. It is important to choose properly the solvents and antisolvents for the formation of LNP; good miscibility between the lignin and the solvent is the best choice. Solvents with good miscibility and a low boiling point lead to the formation of smaller size LNP and with a spherical shape, but if the solvent has poor miscibility and the antisolvent has a higher boiling point, the resulting LNPs do not have a uniform shape and are larger [58]. But other parameters such as the lignin concentration in the solvent, antisolvent temperature, and stirring rate are also important [58].

pH shifting—this technique has similarities with the posterior technique; but here the pH is changed from acid to base or vice versa. Lignin solubility changes with the pH of the solution, and precipitates. As examples of these methods: (i) the lignin is dissolved in NaOH at pH 12 and by HNO_3 the LNPs precipitate [102]; (ii) the lignin is dissolved in ethylene glycol and the LNPs are precipitated by adding hydrochloric acid [95]. The use of an acid to precipitate the LNP makes the process more expensive and dangerous, but the acid can be recovered, and the method has the advantage of attaining well-defined LNP with a good shape and stability [58].

Solvent exchange or solvent shifting—this technique involves a dual system of solvents, an organic solvent (where the lignin is dissolved), and water (in excess allowing the formation of LNP due to the decrease of lignin solubility in water at pH neutral). The LNP size ranges from nanometer to hundreds of micrometers with spherical-shaped particles [58]. Suitable solvents are: THF, acetone, acetone/water, and DMSO [58]. This is similar to the antisolvent precipitation methods, but it is the excess of water that makes the LNP; in the other method, it was the pH shifting. For the solvent exchange method, the dissolved lignin is put in dialysis bags immersed in water and the LNPs are formed during the dialysis process. LNPs were produced by dialysis of a mixture of lignin solution and oleic acid coated with Fe_3O_4 [62]. The need for dialysis bags increases the costs of this method (**Table 8**).

Supercritical fluid (SCF)—this is a technique widely used in the pharmaceutical industry. Generally, the supercritical fluid used is CO_2 due to its diverse advantages (abundance, nontoxicity, low cost, and not flammable). This technique has different names depending on the conditions and the fluids used: supercritical antisolvent (SAS), supercritical gas antisolvent (GAS) are just two examples; LNPs were prepared by SAS, using acetone (solvent) and CO_2 (antisolvent) from poplar organosolv lignin; because the solubility was enhanced, the LNP formed (0.144 μm) presented higher antioxidant capacities compared with the starting lignin [103].

Method	Advantages & disadvantages
Antisolvent precipitation	**Advantages:** good lignin dissolution; low-cost solvents; LNP synthesis with adequate size and morphology. **Disadvantages:** some solvents are expensive, and others are toxic or difficult to recover.
Solvent exchange or shifting	**Advantages:** high lignin dissolution; some solvents can be recovered (in a rotavapor almost 95% of acetone can be recovered), but others are more difficult (e.g. DMSO, high boiling point (189°C)); high yield of LNP production; use of water as antisolvent ⇒ low cost. **Disadvantages:** high consumption of solvent ⇒ high cost; choosing the best affordable solvent is essential. Recovery is difficult; dialysis bags are expensive. Low yield of LNP, and difficulty in controlling their morphology.
Supercritical fluid	**Advantages:** industrial scale-up; able to control the size and shape of the nanoparticles easily (by changing the parameters); high yields. **Disadvantages:** lignin solubility in supercritical CO_2 ⇒ add methanol or ethanol to overcome this; need expertise to run with the equipment, sophisticated laboratory facilities; expensive maintenance.
Ultrasonication	**Advantages:** fast, easy and green synthesis of LNP, low requirements of equipment ⇒ low cost; is harmless to the environment. **Disadvantages:** low yield of LNP, particles with irregular morphology and shape.
Interfacial crosslinking, Polymerization and Nanoprecipitation	**Advantages:** good lignin dissolution; LNP produced with desired morphology and sizes. **Disadvantages:** use expensive and toxic solvents (epichlorohydrin, hydrogen peroxide); removability of the templates is an issue.
Aerosol flow	**Advantages:** one step, high yields and no need for antisolvent. **Disadvantages:** use expensive solvents and some are toxic.
Electrospinning	**Disadvantages:** associated with physicochemical properties and the solubility of lignin.
Biological pathway	**Advantages:** simple and low-cost production of LNP; use of green solvents. **Disadvantages:** low yield of LNP production and with irregular morphology and shape (usual macro).

Table 8.
Some advantages and disadvantages of the methods used for the synthesis of LNP. Adapted from Schneider et al. [60] and Kumar et al. [58].

Aerosol process—this is a single-step method for the production of nanoparticles applied in biomedical applications and was first presented by Eerikäinen et al. [104]; it consists of an atomizer that works continuously to generate an aerosol. This aerosol passes through a heated tube connected to a low-pressure impactor where the particles formed are collected. Researchers were successfully able to produce LNPs with sizes ranging from 30 to 2000 nm [105].

Ultrasonication or acoustic cavitation—the lignin is an aqueous solution and this suspension is sonicated by changing the intensity, time, and temperature; then the sample is dried under mild conditions [106]. These authors used two commercial alkali lignins suspensions to produce efficiently nanosized particles (spherical shapes and homogeneous). If this technique is used alone, the LNPs formed are not uniform in terms of size or the particle size distribution, which are dependent on the ultrasound conditions. Some researchers have overcome this by producing the LNP by solvent exchange combined with ultrasound. This combination is advantageous because the LNPs are rapidly formed and easily detached by centrifugation.

Electrospinning—this is a popular technique for the production of nanomaterials; the lignin is dissolved in a suitable solvent and the mixture is put in a syringe, pumped through a nozzle with reduced size (diameter of around 100 μm) that jets the liquid into a collector plate, creating a nanofiber matrix; an electric field is used with voltage of 100–500 kV/m to create the nanofibers (10–25 cm). Technical lignins (softwood Kraft, hardwood Kraft, and sulfonated Kraft lignins) were mixed with PEO (polyethylene oxide) at different percentages (1–5 wt%) and obtained fibers by electrospinning [107]; the incorporation of PEO improved the fibers produced (more uniformity) at lower viscosities. However, this technique has limitations associated with the lignin properties and it was suggested that electrospraying would be more favorable for nanoparticles production [58].

Interfacial crosslinking—this is not a method for LNP, but is better for their stabilization, generating particles and capsules. The process involves lignin emulsification, usually in an oil-water phase, where the addition of a cross-linking agent improves the lignin cross-link. First, the lignin is dissolved in an alkaline solution that will activate the OH chains and generate a dispersed phase; or it will be dissolved in a solvent with a surfactant such as Span80®, 1-pentanol, or Tween 80®. An oil/water microemulsion is prepared and emulsified to the solubilized lignin, creating the lignin cross-linkages with the water/oil interface. The next step is to add the cross-linking agent, which can be toluene diisocyanate, thiol, or epichlorohydrin [60].

Polymerization—this is similar to the cross-linking method. The lignin has to be modified or some groups activated so that they can be grafted onto a surface that has an affinity to the lignin. As examples of the application of this technique: lignin nanotubes were produced in an alumina membrane that was dissolved in phosphoric acid [108]; lignin nanoparticles were produced by lignin being grafted onto PDMAEMA (2-(dimethylamino)ethyl-methacrylate) and PDEAEMA (2-(diethylamino)ethyl methacrylate) [109, 110].

Biological pathway—this is not a common technique for LNP synthesis, so far. LNPs are produced by microorganisms or by enzymes (formed by either bacteria or fungi). Examples: (i) enzymes were used to break the linkages between lignin and cellulose in *Luffa* fibers, producing lignin-derived particles with cuboidal shape and size around 20–100 nm [111]; (ii) lignin was extracted by coconut fibers (soda pulping), the lignin obtained was hydrolyzed using *Aspergillus* sp. Nanolignin was synthetized by different techniques at distinct yields: homogenization (81%), ultrasonication (64.3%), and microbial (58.4%) [76]; (iii) the colloidal stability of LNP was improved using fungi species [87].

Table 8 presents some of the methods mentioned and their strengths, as well as some of their weaknesses. The comparison was made regarding the type of solvents used (if environmentally-safe, cost), but also regarding the process time, LNP morphology, and size. Generally, solvents such as acetone, dioxane, THF, and DMSO are used in medium volume, purchased, and are relatively affordable. The use of hazardous reagents is a negative point. Some studies include green solvents such as levulinic acid that had high efficiency in dissolving Kraft lignin (up to 40%) compared with other carboxylic acids, and the lignin dissolved can be regenerated by adding an

Method	Method resume	Morphology & size	Potential applications	Ref
Antioxidants				
Antisolvent	• Lignin dissolved in acetone: water (9:1, v/v) • Dissolved lignin precipitate in water	Spheres 104 nm	Properties: antioxidant and UV-protection ⇒ food, pharmaceutical, cosmetic industries	[73]
Supercritical antisolvent (SAS)	• Organosolv lignin used to prepare nanoscale lignin by antisolvent precipitation using acetone (solvent) and supercritical carbon dioxide (antisolvent) • SAS did not degrade the lignin, but increased its solubility (12.4 times) ⇒ enhanced the antioxidant properties and free radical scavenging activities	Spheres 0.144 μm	Nanoscale lignin presented higher antioxidant activities than starting lignin ⇒ animal husbandry, food processing and pharmaceutical industries	[103]
Antisolvent	• Lignin dissolved in ethylene glycol • Precipitation in HCl	181 nm	Reinforcing agent in polymers and nanocomposites	[119]
Nanocomposites				
Self-assembly	• Alkali lignin (100 g) dissolved in dioxane (1 L) • Adding cyclohexane (100 mL) ⇒ flocculation and precipitation of lignin (formation of lignin reverse micelles—LRM, separated and dried by vacuum)	Micelles 120 nm to micrometers	Properties: UV-absorbance and good miscibility with plastics such as high-density polyethylene (HDPE) ⇒ suitable for blends with HDPE, improving mechanical properties	[120]
Aerosol flow reactor	• Different lignins: Kraft (KL), organosolv (OS) and alkali lignin (AL) • AL dissolved in water, KL and OS dissolved in dimethylformamide (DMF) – concentrations 0.5–2% • Drops were generated in a Collison-type jet atomizer with nitrogen gas, carried to a heated laminar flow reactor, collected as solid particles and dried • Solid particles were fractionated with a Berner-type low-pressure impactor ⇒ size fractionation (30 nm to 2 μm)	Spheres 30 nm–2 μm	Stabilizes oil-in-water Pickering emulsions	[105]
Electrospinning	• Preparation of polyethylene oxide solution with acetonitrile (10% vol) and sodium hydroxide (0.1 N) stirring overnight; then addition of soy protein	Nanofibers 124–400 nm	Composite development	[121]

Method	Method resume	Morphology & size	Potential applications	Ref
	under stirring (4 h) until complete dissolution; only then Kraft lignin in powder was added slowly; stirring further for 4 h • Polyethylene oxide (coadjutant) was used to facilitate the formation of nanofibers with high quality			
Delivery systems/encapsulation				
Antisolvent	• Nanoparticles (NP) production by two methods: • Low-sulfonated lignin (Indulin® AT, IAT, 0.28 g) dissolved in ethylene glycol (ET, 50 mL) and precipitation by adding diluted HCl (8 mL, 0.025 M) ⇒ yield NP stable at wide range of pH (1–9). NP stability due to the stacking of uncharged lignin in ET upon precipitation • IAT dissolved in high-pH aqueous solution (NaOH, 1 N, pH 12) and acidic precipitation with HNO_3 (0.25 M, 900 μL) ⇒ NP stable at low pH (dissolve at pH > 5)	100 nm to micrometers	NP highly porous ⇒ promising vehicles for drug delivery, stabilizers of cosmetics and pharmaceutical preparations, and sorbents for heavy metal ions and other pollutants	[98]
Antisolvent	• Indulin AT lignin dissolved in ethylene glycol and precipitation in HCl • LNP infused with silver ions and coated with poly-(diallyldimethylammonium-chloride- PDAC) to boost the adhesion to microbial cell membranes		LNP infused with silver ions and coated with cationic polyelectrolyte (PDAC) layer ⇒ synthesis of biodegradable cores that have high antimicrobial activity	[102]
Self-assembly	• Alkali lignin was acetylated, dissolved in tetrahydrofuran (THF) • LNP formation by adding water, then the THF was evaporated in rotavapor to attain colloidal spheres	Colloidal lignin spheres	LNP for drug delivery, microencapsulation of pesticides	[122]
Interfacial crosslinking	• Lignin dissolved with sodium chloride in water, mixed with cyclohexane containing polyglycerol polyricinoleate (surfactant); • Ultrasonification to produce an emulsion by adding toluene diisocyanate (TDI)	Hollow nanocapsules 150–200 nm	LNP encapsulated with hydrophilic compounds (e.g. fertilizers, pesticides and drugs), that can be released by enzymatic degradation ⇒ LNP suitable for agricultural purposes	[123]

Method	Method resume	Morphology & size	Potential applications	Ref
Interfacial crosslinking	• Creation of an emulsion of water, oil and lignin • Ultrasonic technology to create cross-linkages between lignin and the water/oil interface	Sphere lignin micro (LMC) and nanocapsules (LNC) 300 nm to 1 μm	Lignin capsules (LMC/ LNC) encapsulated with oil ⇒ potential use for biomedical purposes	[66]
Aerosol assisted self-assembly	• Alkali lignin was dissolved in dioxane, dried in rotavapor (after removal of insoluble fraction), and dissolved in DMSO at different concentrations • DMSO solutions were sprayed by ultrasonic nebulizer dispersed in liquid nitrogen upon a cooled Cu-plate and collected as frozen drops	Hollow solid lignin colloidal nanoparticles without chemical modifications 96–193 nm	LNP have UV absorbing ability ⇒ sunscreen cosmetics; drug delivery application	[84]
Polymerization	• Lignins isolated by different methods: sulfuric acid (Klason lignin), thioglycolic acid, phosphoric acid, sodium hydroxide • Lignins reacted with amine groups on the nanopores surface of the alumina membrane template ⇒ dehydrogenation of lignin • Deposition of multiple layers of lignin monomers onto base layer • Dissolution of alumina templates in phosphoric acid to obtain lignin nanotubes • LNT can be tailored according to the application	Lignin nanotubes (LNT)	DNA can adsorb to LNT ⇒ useful for genetic delivery into human cells. LNT from NaOH-extracted lignin were quite interesting (short in length, nucleus easily penetrated ⇒ enhance the delivery of DNA)	[108]
Self-assembly	• Kraft lignin (KL, 1 g) dissolved in ethanol (ET, 50 mL) and then ultrasound to accelerate the dissolution (2–3 min), site for 72 h ⇒ insoluble separated by centrifugation, supernatant filtered by 0.45 μm syringe • LNC formed by adding water to the solution KL/ET	KL nanocapsules (KLC)	KLC can be applied in a wide range of applications	[124]
Biological pathway	• Dried fruit of *Luffa cylindrica* was extracted with acetone, ethanol and tholuene (1:1:4) for 6 h in soxhlet apparatus • Extractive-free fibers were milled and soaked in acetate	LNP cuboidal shape 20–100 nm	LNP potential applications for wide range of fields from pharmaceutical, automobile and polymer industries	[111]

Method	Method resume	Morphology & size	Potential applications	Ref
	buffer (pH 4.5, 1–2 h) and were tripled enzymatic treated with MultifectCX GC Cellulase (10%), Multifect Pectinase FE (15%), Optimase CX 255 L (15%) to produce cuboidal lignin nanoparticles			
Antisolvent	• Wheat straw alkali lignin, without any modification was dissolved by ionic liquids (e.g. [Emim][Ac] at concentration 5– 200 mg/mL, stirred at room temperature) • Water added (speed 2–6 mL/min to 80%) to produce lignin nanospheres stable at pH between 4 and 10 • LNS were freeze-dried and stored; LNS were stable in water without aggregation within 60 days (stability attributed to an electrical double layer formed by the phenolic OH and possible carboxylic groups that repulse nanospheres from each other)	Lignin nanospheres (LNS)	LNS with potential for coatings, drug delivery and nanocomposites production	[117]
Self-assembly	• Isolation of lignin from corncob residues by choline chloride/lactic acid-based deep eutectic solvent (DES) • Lignin was dissolved in acetone/water solution (4:1, v/v), filtrated to remove undissolved lignin • Water was added (antisolvent) under moderate stirring ⇒ LNP production with high stability	LNP sphere structure, with small size, well-defined shape 48.66–183.7 nm		[43]

Table 9.
Compilation of some potential uses and methods for lignin nanoparticles production.

excess of water, maintaining the morphological characteristics and the thermal stability of the lignin [112].

Researchers are able to produce LNP with a range of shapes, such as nanoparticles [62–64], nanotubes [108], nanospheres [113], nanofibers [114], nanocapsules [115], nanomicelles [85], nanofilms [116]; and sizes (micrometers and nanometers). This high diversity of shapes and sizes contrasts with a high demand for uniformity for specific applications, which requires reproducibility at industrial-scale production [58]. According to Zhang et al. [43], within a certain lignin concentration range, the particle size increased with increasing lignin concentration. A higher concentration of lignin in the system means more lignin available for the growth of the nanospheres (LNS); a high stirring rate of aqueous phase and ionic liquids produces smaller particles. Also, the dropping speed of water affects the diameters of the LNS, where the increase of the water dropping speed from 2 to 6 mL/s decreased the LNS diameter [117]. Besides these aspects, the feasibility of LNP production depends on the scale-up, manufacturing costs, and applications. Having this in mind, Assis and coworkers [118] carried out a techno-economic study to evaluate the costs of lignin production, the costs of LNP synthesis, and their applications. **Table 9** presents a list of LNPs produced by different methods and compiled by their possible applications.

5. Conclusions

This chapter includes an overview on lignin and lignin nanoparticles. Lignin is a complex natural polymer with characteristics that contribute to the development and growth of plants. Lignin provides rigidity, hydrophobicity, and protection against biotic agents and plays an important role in the plant's adaptability to the environment. Plants can produce different monomers as a response to stress (e.g., drought, high or low temperature, and light).

As a natural polyphenol polymer, lignin presents high variability associated with the diversity of plants, which creates great challenges for its recovery and valorization.

There are several methods to deconstruct the biomass and obtain the so-called *technical lignin*. These approaches present some advantages, the major one being to obtain a pulp (a valuable product). However, from the lignin point of view, there are several disadvantages: lignin is highly condensed, difficult to recover and process for subsequent valorization. Aiming to change this, a new perspective has been used, the *lignin-first* approach. Here, lignin is obtained first to prevent adverse reactions and be easier to valorize under the scope of the biorefinery. The *lignin-first* approach includes different methods also due to lignin's natural diversity, and it is still hard to attain lignin both in high yield and with good characteristics for different purposes.

The production of lignin nanoparticles (LNPs) started, at the beginning of 2000, because the nano-size improves the lignin characteristics (e.g., increased antioxidant activity). LNPs can be produced by different methods, where a diverse range of sizes and shapes can be obtained and tested for multiple applications. However, there are some aspects that still have to be improved so that LNP can be scaled up (e.g., costs of lignin isolation and LNP production, use of more green solvents).

Undoubtedly, lignin is an unfinished theme for research due to its chemical characteristics and endless potential applications.

Acknowledgements

This work was supported by FCT (Fundação para a Ciência e Tecnologia, Portugal) by financing the: Forest Research Center (UIDB/00239/2020), Acacia4FirePrev (PCIF/GVB/0145/2018) and CleanForest (PCIF/GVB/0167/2018) projects. Ana Lourenço was supported by FCT through a research contract (DL 57/2016/CP1382/CT0007). The authors also acknowledge the COST Action CA17128 "Establishment of a Pan-European Network on the Sustainable Valorisation of Lignin (LignoCOST, https://lignocost.eu/ (accessed on September 30, 2022))", supported by COST (European Cooperation in Science and Technology) for promoting exchange and dissemination of knowledge/expertise between researchers from different countries.

Conflict of interest

The authors declare no conflict of interest.

Author details

Ana Lourenço[1,2]* and Jorge Gominho[1,2]

1 Forest Research Centre, School of Agriculture, University of Lisbon, Lisbon, Portugal

2 Associate Laboratory Terra, School of Agriculture, University of Lisbon, Portugal

*Address all correspondence to: analourenco@isa.ulisboa.pt

References

[1] Chatterjee S, Saito T. Solvent fractionation of lignin. Chapter 7. In: Naskar AK, Hoffman WP, editors. Polymer Precurso-derived carbon, ACS Symposium Series. Washington: American Chemical Society; 2014. DOI: 10.1021/bk-2014-1173

[2] Bonawitz ND, Chapple. The genetics of lignin biosynthesis: Connecting genotype to phenotype. Annual Review of Genetics. 2010;**44**:337-363. DOI: 10.1146/annurev-genet-102209-163508

[3] Lourenço A, Pereira H. Compositional variability of lignin in biomass. In: Poletto M, editor. Chapter 3 of the Book "Lignin – Trends and Applications". InTechOpen; 2018. pp. 65-98. DOI: 10.5772/intechopen.71208. Available from: https://www.intech open.com/books/lignin-trends-and-applications

[4] Lourenço A, Rencoret J, Chemetova C, Gominho J, Gutiérrez A, del Río JC, et al. Lignin composition and structure differs between xylem, phloem and phellem in *Quercus suber* L. Frontiers in Plant Science. 2016;7:1612. DOI: 10.3389/fpls.2016.01612

[5] Katahira R, Elder T, Beckham GT. A brief introduction to lignin structure. Chapter 1. In: Beckham GT, editor. Lignin Valorization. Emerging Approaches. UK: Royal Society of Chemistry, Energy and Environmental Series N° 9; 2018. pp. 1-15. ISBN: 978-1-78262-554-4

[6] Lourenço A, Araújo S, Gominho J, Pereira H, Evtuguin D. Structural changes in lignin of thermally treated eucalyptus wood. Journal of Wood Chemistry and Technology. 2020;**40**(4): 258-268. DOI: 10.1080/02773813.2020. 1769674

[7] Abu-Omar MM, Barta K, Beckhman GT, Luterbacher JS, Ralph J, Rinaldi R, et al. Guidelines for performing lignin-first biorefining. Energy & Environmental Science. 2021; **14**:262. DOI: 10.1039/D0EE02870C

[8] Schutyser W, Renders T, Bosch AV, Koelewijin SF, Beckham GT, Sels BF. Chemicals from lignin: An interplay of lignocellulose fractionation, depolymerization, and upgrading. Chemical Society Reviews. 2018;**47**(3): 852-908. DOI: 10.1039/C7CS00566K

[9] Ralph J. Hydroxycinnamates in lignification. Phytochemistry Reviews. 2010;**9**:65-83. DOI: 10.1007/s11101-009-9141-9

[10] Del Río JC, Rencoret J, Prinsen P, Martínez ÁT, Ralph J, Gutiérrez A. Structural characterization of wheat straw lignin as revealed by analytical pyrolysis, 2D-NMR, and reductive cleavage methods. Journal of Agricultural and Food Chemistry. 2012; **60**:5922-5935. DOI: 10.1021/jf301002n

[11] Mottiar Y, Vanholme R, Boerjan W, Ralph J, Mansfield SD. Designer lignins: Harnessing the plasticity of lignification. Current Opinion in Biotechnology. 2016; **37**:190-200. DOI: 10.1016/j. copbio.2015.10.009

[12] Singh S, Bashri G, Singh A, Prasad SM. Regulation of Xenobiotics in Higher Plants: Signalling and Detoxification. Singapore: Springer; 2016. pp. 39-56. Available from: https:// link.springer.com/chapter/10.1007/ 978-981-10-2860-1_3

[13] Lan W, Lu F, Regner M, Zhu Y, Rencoret J, Ralph SA, et al. Tricin, a flavonoid monomer in monocot

lignification. Plant Physiology. 2015; **167**(4):1284-1295. DOI: 10.1104/pp.114.253757

[14] Eloy NB, Voorend W, Lan W, Saleme MLS, Cesarino I, Vanholme R, et al. Silencing chalcone synthase impedes the incorporation of tricin in lignin and increases lignin content. Plant Physiology. 2017;**173**(2):998-1016. DOI: 10.1104/pp.16.01108

[15] Del Río JC, Rencoret J, Gutiérrez A, Kim H, Ralph J. Hydroxystilbenes are monomers in palm fruit endocarp lignins. Plant Physiology. 2017;**174**: 2072-2082. DOI: 10.1104/pp.17.00362

[16] Feghali E, van de Pas DJ, Parrott AJ, Torr KM. Biobased epoxy thermoset polymers from depolymerized native hardwood lignin. ACS Macro Letters. 2020;**9**(8):1155-1160. DOI: 10.1021/acsmacrolett.0c00424

[17] Questello-Santiago YM, Galkin MV, Barta K, Luetrbacher JS. Stabilization strategies in biomass depolymerization using chemical functionalization. Nature Reviews Chemistry. 2020;**4**:311-330. DOI: 10.1038/s41570-020-0187-y

[18] Liu X, Bouxin FP, Fan J, Budarin VL, Hu C, Clark JH. Recent advances in the catalytic depolymerization of lignin towards phenolic chemicals: A review. ChemSusChem. 2020;**13**:4296-4317. DOI: 10.1002/cssc.202001213

[19] Boudet A. Lignins and lignification: Selected issues. Plant Physiology and Biochemistry. 2000;**38**(1/2):81-96. DOI: 10.1016/S0981-9428(00)00166-2

[20] Davin LB, Lewis NG. Lignin primary structures and dirigent sites. Current Opinion in Biotechnology. 2005;**16**: 407-415. DOI: 10.1016/j.copbio.2005.06.011

[21] Liu Q, Luo L, Zheng L. Lignins: Biosynthesis and biological functions in plants. International Journal of Molecular Sciences. 2018;**19**(2):335. DOI: 10.3390/ijms19020335

[22] Khan P, Tong L, Khan SU, Zhang C, Wang W. Lignin: A defensive shield halting the environmental stresses—A review. Applied Ecology and Environmental Research. 2022;**20**(3): 1991-2015. DOI: 10.15666/aeer/2003_19912015

[23] Cesarino I. Structural features and regulation of lignin deposited upon biotic and abiotic stresses. Current Opinion in Biotechnology. 2019;**56**: 209-214. DOI: 10.1016/j.copbio.2018.12.012

[24] Hawkins S, Boudet A. Defense lignin and hydroxycinnamyl alcohol dehydrogenase activities in wounded *Eucalyptus gunnii*. Forest Pathology. 2003;**33**(5):339-352. DOI: 10.1046/j.1439-0329.2003.00341.x

[25] Moura JCMS, Bonine CAV, Viana JOF, Dornelas MC, Mazzafera P. Abiotic and biotic stresses and changes in the lignin content and composition in plants. Journal of Integrative Plant Biology. 2010;**52**(4):360-376. DOI: 10.1111/j.1744-7909.2010.00892.x

[26] Han X, Zhao Y, Chan Y, Xu J, Jiang C, Wang X, et al. Lignin biosynthesis and accumulation in response to abiotic stresses in woody plants. Forestry Research. 2022;**2**:9. DOI: 10.48130/FR-2022-0009

[27] Bhardwaj R, Handa N, Sharma R, Kaur H, Kohli S, Kumar V, et al. Lignins and abiotic stress: An overview. In: Ahmad P, Wani MR, editors. Chapter in "Physiological Mechanisms and Adaptation Strategies in Plants under Changing Environment". New York:

Springer Science + Business Media; 2013. pp. 267-296. DOI: 10.1007/978-1-4614-8591-9_10

[28] Gall H, Phillippe F, Domon J, Gillet F, Pelloux J, Rayoon C. Cell wall metabolism in response to abiotic stress. Plants. 2015;**4**:112-166. DOI: 10.3390/plants4010112

[29] Chun HJ, Baek D, Cho HM, Lee SH, Jin BJ, Yun D, et al. Lignin biosynthesis genes play critical roles in the adaptation of Arabidopsis plants to high-salt stress. Plant Signaling & Behavior. 2019;**14**(8): 1625697. DOI: 10.1080/15592324.2019.1625697

[30] Hori C, Yu X, Mortimer JC, Sano R, Matsumoto T, Kikuchi J, et al. Impacto f abiotic stress on the regulation of cell wall biosynthesis in *Populus trichocarpa*. Plant Biotechnology. 2020;**37**(3): 273-283. DOI: 10.5511/plantbiotech nology.20.0326a

[31] Yao T, Feng K, Xie M, Barros J, Tschaplinski TJ, Tuskan GA, et al. Phylogenetic occurrence of the phenylpropanoid pathway and lignin biosynthesis in plants. Frontiers in Plant Science. 2021;**12**:704697. DOI: 10.3389/fpls.2021.704697

[32] Björkman A. Studies on finely divided wood. Part I. extraction of lignin with neutral solvents. Sven Papperstidn. 1956;**13**:477-485

[33] Evtuguin DV, Neto CP, Silva AMS, Domingues PM, Amado FML, Robert D, et al. Comprehensive study on the chemical structure of dioxane lignin from plantation *Eucalyptus globulus* wood. Journal of Agricultural and Food Chemistry. 2001;**49**:4252-4261. DOI: 10.1021/jf010315d

[34] Bozell JJ, Chmely SE, Hartwig W, Key R, Labbé N, Venugopal P, et al. Lignin isolation methodology for biorefining, pretreatment and analysis. In: Beckham GT, editor. Chapter 2 of "Lignin Valorization. Emerging Approaches". UK: Royal Society of Chemistry, Energy & Environment Series Nº 9; 2018. pp. 21-61. ISBN: 978-1-78262-554-4

[35] Gigli M, Crestini C. Fractionation of industrial lignins: Opportunities and challenges. Green Chemistry. 2020;**22**: 4722-4746. DOI: 10.1039/D0GC01606C

[36] Boeriu C, Fitigău FI, Gosselink RJA, Frissen AE, Stoutjesdijk J, Peter F. Fractionation of five technical lignins by selective extraction in green solvents and characterisation of isolated fractions. Industrial Crops and Products. 2014;**62**: 481-490. DOI: 10.1016/j.indcrop.2014.09.019

[37] Naseem A, Tabasum S, Zia KM, Zuber M, Ali M, Noreen A. Lignin-derivatives based polymers, blends and composites: A review. International Journal of Biological Macromolecules. 2016;**93**(A):263-313. DOI: 10.1016/j.ijbiomac.2016.08.030

[38] Ekielski A, Miska PK. Lignin for bioeconomy: The present and future role of technical lignin. International Journal of Molecules Sciences. 2021;**22**:63. DOI: 10.3390/ijms22010063

[39] Lourenço A, Gominho J, Pereira H. Pulping and delignification of sapwood and heartwood from *Eucalyptus globulus*. Journal of Pulp and Paper Science. 2010; **36**(3-4):63-69. DOI: not assigned

[40] Gominho J, Fernandez J, Pereira H. *Cynara cardunculus* L.—A new fibre crop for ulp and paper production. Industrial Crops and Products. 2001;**13**:1-10. DOI: 10.1016/S0926-6690(00)00044-3

[41] Aro T, Fatechi P. Production and application of lignosulfonates and

sulfonated lignin. ChemSusChem. 2017; **10**:1861-1877. DOI: 10.1002/cssc.201700082

[42] Overejo-Pérez A, Rigual V, Domínguez JC, Alonso MV, Oliet M, Rodriguez F. Acidic depolymerization vs ionic liquid solubilization in lignin extraction from eucalyptus wood using the protic ionic liquid 1-methylimidazolium chloride. International Journal of Biological Macromolecules. 2020;**157**:461-469. DOI: 10.1016/j.ijbiomac.2020.04.194

[43] Zhang H, Shi Y, Li M, Chen J, Xin Y, Zhang L, et al. Extraction of lignin from corncob residue via a deep eutectic solvent for the preparation of nanoparticles by self-assembly. Chemical Engineering Science. 2022;**256**: 117694. DOI: 10.1016/j.ces.2022.117694

[44] Korányi T, Fridrich B, Pineda A, Barta K. Development of 'lignin-first' approaches for the valorization of lignocellulosic biomass. Molecules. 2020; **25**:2815. DOI: 10.3390/molecules25122815

[45] Zhou N, Thilakarathna WPDW, He QS, Rupasinghe HPV. A review: Depolymerization of lignin to generate high-value bio-products: Opportunities, challenges and prospects. Frontiers in Energy Research. 2022;**9**:758744. DOI: 10.3389/fenrg.2021.758744

[46] Renders T, Van den Bosch S, Koelewijn S-F, Schutyser W, Sels BF. Lignin-first biomass fractionation: The advent of active stabilisation strategies. Energy Environmental Science. 2017;**10**: 1551-1557. DOI: 10.1039/C7EE01298E

[47] Galkin M. From stabilization strategies to tailor-made lignin macromolecules and oligomers for materials. Current Opinion in Green and Sustainable Chemistry. 2021;**28**:100438. DOI: 10.1016/j.cogsc.2020.100438

[48] Strassberger Z, Prinsen P, Klis FVD, Es DSV, Tanase S, Rothenberg G. Lignin solubilisation and gentle fractionation in liquid ammonia. Green Chemistry. 2015; **17**(1):325-334. DOI: 10.1039/C4GC01143K

[49] Thoresen PP, Matsakas L, Rova U, Christakopoulos P. Recent advances in organosolv fractionation: Towards biomass fractionation technology of the future. Bioresources Technology. 2020; **306**:123189. DOI: 10.1016/j.biortech.2020.123189

[50] Ahmed MA, Lee JH, Raja AA, Choi JW. Effects of gamma-valerolactone assisted fractionation of ball-milled pine wood on lignin extraction and its characterization as well as its corresponding cellulose digestion. Applied Sciences. 2020;**10**(5): 1599. DOI: 10.3390/app10051599

[51] Xu J, Dai L, Gui Y, Yuan L, Zhang C, Lei Y. Synergistic benefits from a lignin-first biorefinery of poplar via coupling acesulfamate ionic liquid followed by mild alkaline extraction. Bioresource Technology. 2020;**303**:122888. DOI: 10.1016/j.biortech.2020.122888

[52] Shuai L, Amiri MT, Questell-Santiago YM, Héroguel F, Li Y, Kim H, et al. Formaldehyde stabilization facilitates lignin monomer production during biomass depolymerization. Science. 2016;**354**(6310):329-333. Available from: https://www.science.org/doi/10.1126/science.aaf7810

[53] Paone E, Tabanelli T, Mauriello F. The rise of lignin biorefinery. Current Opinion in Green Sustainable Chemistry. 2020;**24**: 1-6. DOI: 10.1016/j.cogsc.2019.11.004

[54] Karlsson M, Giummarella N, Lindén PA, Lawoko M. Toward a consolidated lignin biorefinery: Preserving the lignin structure through

additive-free protection strategies. ChemSusChem. 2020;**13**:4666-4677. DOI: 10.1002/cssc.202000974

[55] Xu J, Zhou P, Zhang C, Yuan L, Xiao X, Dai L, et al. Striding the threshold of photocatalytic lignin-first biorefining via a bottom-up approach: From model compounds to realistic lignin. Green Chemistry. 2022; **24**(14):5351-5378. DOI: 10.1039/D2GC01409B

[56] Sadeghifar H, Ragasukas A. Perspective on technical lignin fractionation. ACS Sustainable Chemistry & Engineering. 2020;**8**(22): 8086-8101. DOI: 10.1021/acssuschemeng.0c01348

[57] Olivares M, Guzmán JA, Natho A, Saavedra A. Kraft lignin utilization in adhesives. Wood Science and Technology. 1988;**22**:157-165. DOI: 10.1007/BF00355851

[58] Kumar R, Gupta A, Chawla M, Aadil KR, Dutt S, Kumar VB, et al. Advances in nanotechnology based strategies for synthesis of nanoparticles of lignin. In: Sharma S, Kumar A, editors. Lignin. Biosynthesis and Transformation for Industrial Applications. Switzerland AG: Springer Series on Polymer and Composite Materials. 2020. pp. 203-229

[59] Becker J, Wittmann C. A field of dreams: Lignin valorization into chemicals, materials, fuels, and health-care products. Biotechnology Advances. 2019;**37**(6):107360. DOI: 10.1016/j.biotechadv.2019.02.016

[60] Schneider WDH, Dillon AJP, Camassola M. Lignin nanoparticles enter the scene: A promising versatile green tool for multiple applications. Biotechnology Advances. 2021;**47**: 107685. DOI: 10.1016/j.biotechadv.2020.107685

[61] Lora J. Industrial commercial lignins: Sources, properties and applications. In: Belgacem M, Gandini A, editors. Monomers, Polymers and Composites from Renewable Resources. Oxford, UK: Elsevier; 2008. pp. 225-241

[62] Figueiredo P, Lintinen K, Kiriazis A, Hynninen V, Liu Z, Bauleth-Ramos T, et al. In vitro evaluation of biodegradable lignin-based nanoparticles for drug delivery and enhanced antiproliferation effect in cancer cells. Biomaterials. 2017;**121**: 97-108. DOI: 10.1016/j.biomaterials.2016.12.034-2017a

[63] Figueiredo P, Ferro C, Kemell M, Liu Z, Kiriazis A, Lintinen K, et al. Functionalization of carboxylated lignin nanoparticles for targeted and PH-responsive delivery of anticancer drugs. Nanomedicine. 2017;**12**: 2581-2596

[64] Figueiredo P, Lintinen K, Hirvonen JT, Kostiainen MA, Santos HA. Properties and chemical modifications of lignin: Towards lignin-based nanomaterials for biomedical applications. Progress in Materials Science. 2018;**93**:233-269. DOI: 10.1016/j.pmatsci.2017.12.001

[65] Mattos BD, Tardy BL, Magalhães WLE, Rojas OJ. Controlled release for crop and wood protection. Recent progress toward sustainable and safe nanostructured biocidal systems. Journal of Controlled Release. 2017;**262**: 139-150. DOI: 10.1016/j.jconrel.2017.07.025

[66] Tortora M, Cavalieri F, Mosesso P, Ciaffardini F, Melone F, Crestini C. Ultrasound driven assembly of lignin into microcapsules for storage and delivery of hydrophobic molecules. Biomacromolecules. 2014;**15**(5): 1634-1643. DOI: 10.1021/bm500015j

[67] Michelin M, Marques AM, Pastrana LM, Teixeira JA, Cerqueira MA. Carboxymethyl cellulose-based films: Effect of organosolv lignin incorporation on physicochemical and antioxidant properties. Journal of Food Engineering. 2020;**285**:110107. DOI: 10.1016/j.jfoodeng.2020.110107

[68] Iravani S, Varma RS. Greener synthesis of lignin nanoparticles and their applications. Green Chemistry. 2020;**22**:612. DOI: 10.1039/C9GC02835H

[69] Zikeli F, Vinciguerra V, D'Annibale A, Capitani D, Romagnoli M, Scarascia MG. Preparation of lignin nanoparticles from wood waste for wood surface treatment. Nanomaterials. 2019;**9**(2):281. DOI: 10.3390/nano9020281

[70] Nair SS, Sharma S, Pu Y, Sun Q, Pan S, Zhu JY, et al. High shear homogenization of lignin to nanolignin and thermal stability of nanolignin-polyvinyl alcohol blends. ChemSusChem. 2014;**7**(12):3513-3520. DOI: 10.1002/cssc.201402314

[71] Beisl S, Friedl A, Miltner A. Lignin from micro- to nanosize: Applications. International Journal of Molecular Sciences. 2017;**18**(11):2367. DOI: 10.3390/ijms18112367

[72] Pereira AES, Oliveira JL, Savassa SM, Rogério CB, Medeiros GA, Fraceto LF. Lignin naoparticles: Insights for a sustainable agriculture. Journal of Cleaner Production. 2022;**345**:131145. DOI: 10.1016/j.jclepro.2022.131145

[73] Xing Q, Buono P, Ruch D, Dubois P, Wu L, Wang W-J. Biodegradable UV-blocking films through core–shell lignin–melanin nanoparticles in poly (butylene adipate-Co-terephthalate). ACS Sustainable Chemistry & Engineering. 2019;**7**(4):4147-4157. DOI: 10.1021/acssuschemeng.8b05755

[74] Xiao D, Ding W, Zhang J, Ge Y, Wu Z, Li Z. Fabrication of a versatile lignin-based nano-trap for heavy metal ion capture and bacterial inhibition. Chemical Engineering Journal. 2019;**358**:310-320. DOI: 10.1016/j.cej.2018.10.037

[75] Yang W, Fortunati E, Bertoglio F, Owczarek JS, Bruni G, Kozanecki M, et al. Polyvinyl alcohol/chitosan hydrogels with enhanced antioxidant and antibacterial properties induced by lignin nanoparticles. Carbohydrate Polymers. 2018;**181**:275-284. DOI: 10.1016/j.carbpol.2017.10.084

[76] Juikar SJ, Vigneshwaran N. Extraction of nanolignin from coconut fibers by controlled microbial hydrolysis. Industrial Crops and Products. 2017;**109**:420-425. DOI: 10.1016/j.indcrop.2017.08.067

[77] Yang W, Rallini M, Wang D-Y, Gao D, Dominici F, Torre L, et al. Role of lignin nanoparticles in UV resistance, thermal and mechanical performance of PMMA nanocomposites prepared by a combined free-radical graft polymerization/masterbatch procedure. Composites Part A: Applied Science and Manufacturing. 2018;**107**:61-69. DOI: 10.1016/j.compositesa.2017.12.030

[78] Gutiérrez-Hernández JM, Escalante A, Murillo-Vázquez RN, Delgado E, González FJ, Toríz G. Use of *agave Tequilana*-lignin and zinc oxide nanoparticles for skin photoprotection. Journal of Photochemistry and Photobiology B: Biology. 2016;**163**:156-161. DOI: 10.1016/j.jphotobiol.2016.08.027

[79] Wang B, Sun D, Wang H-M, Yuan T-Q, Sun R-C. Green and facile preparation of regular lignin nanoparticles with high yield and their natural broad-spectrum sunscreens. ACS Sustainable Chemistry & Engineering.

2019;7(2):2658-2666. DOI: 10.1021/acssuschemeng.8b05735

[80] Yin H, Liu L, Wang X, Wang T, Zhou Y, Liu B, et al. A novel flocculant prepared by lignin nanoparticles-gelatin complex from switchgrass for the capture of *Staphylococcus aureus* and *Escherichia coli*. Colloids and Surfaces A: Physicochemical Engineering Aspects. 2018;**545**:51-59. DOI: 10.1016/j.colsurfa.2018.02.033

[81] Azimvand J, Didehban K, Mirshokraie S. Safranin-O removal from aqueous solutions using lignin nanoparticle-g-polyacrylic acid adsorbent: Synthesis, properties, and application. Adsorption Science & Technology. 2018;**36**:1422-1440. DOI: 10.1177/0263617418777

[82] Dai L, Liu R, Hu L-Q, Zou Z-F, Si C-L. Lignin nanoparticle as a novel green carrier for the efficient delivery of resveratrol. ACS Sustainable Chemistry & Engineering. 2017;**5**:8241-8249. DOI: 10.1021/acssuschemeng.7b01903

[83] Li Y, Qiu X, Qian Y, Xiong W, Yang D. PH-responsive lignin-based complex micelles: Preparation, characterization and application in oral drug delivery. Chemical Engineering Journal. 2017;**327**:1176-1183. DOI: 10.1016/j.cej.2017.07.022

[84] Mishra PK, Wimmer R. Aerosol assisted self-assembly as a route to synthesize solid and hollow spherical lignin colloids and its utilization in layer by layer deposition. Ultrasonic Sonochemistry. 2017;**35**:45-50. DOI: 10.1016/j.ultsonch.2016.09.001

[85] Sipponen MH, Smyth M, Leskinen T, Johansson L-S, Österberg M. All-lignin approach to prepare cationic colloidal lignin particles: Stabilization of durable Pickering emulsions. Green Chemistry. 2017;**19**:5831-5840. DOI: 10.1039/C7GC02900D

[86] Sipponen MH, Farooq M, Koivisto J, Pellis A, Seitsonen J, Österberg M. Spatially confined lignin nanospheres for biocatalytic ester synthesis in aqueous media. Nature Communications. 2018;**9**:2300. DOI: 10.1038/s41467-018-04715-6

[87] Mattinen M-L, Valle-Delgado JJ, Leskinen T, Anttila T, Riviere G, Sipponen M, et al. Enzymatically and chemically oxidized lignin nanoparticles for biomaterial applications. Enzyme and Microbial Technology. 2018;**111**:48-56. DOI: 10.1016/j.enzmictec.2018.01.005

[88] Mattinen M-L, Riviere G, Henn A, Nugroho RWN, Leskinen T, Nivala O, et al. Colloidal lignin particles as adhesives for soft materials. Nanomaterials. 2018;**8**:1001. DOI: 10.3390/nano8121001

[89] Gonzalez G, Nelly M, Levi M, Turri S, Griffini G. Lignin nanoparticles by ultrasonication and their incorporation in waterborne polymer nanocomposites. Journal of Applied Polymer Science. 2017;**134**:45318. DOI: 10.1002/app.45318

[90] Silmore KS, Gupta C, Washburn NR. Tunable Pickering emulsions with polymer-grafted lignin nanoparticles (PGLNs). Journal of Colloid and Interface Science. 2016;**466**:91-100. DOI: 10.1016/j.jcis.2015.11.042

[91] Tian D, Hu J, Bao J, Chandra RP, Saddler JN, Lu C. Lignin valorization: Lignin nanoparticles as high-value bio-additive for multifunctional nanocomposites. Biotechnology for Biofuels and Bioproducts. 2017;**10**:192. DOI: 10.1186/s13068-017-0876-z

[92] Tian D, Hu J, Chandra RP, Saddler JN, Lu C. Valorizing recalcitrant cellulolytic enzyme lignin via lignin nanoparticles fabrication in an integrated biorefinery. ACS Sustainable Chemistry & Engineering. 2017;**5**: 2702-2710. DOI: 10.1021/acssuschemeng.6b03043

[93] Gong W, Ran Z, Ye F, Zhao G. Lignin from bamboo shoot shells as an activator and novel immobilizing support for -amylase. Food Chemistry. 2017;**228**:455-462. DOI: 10.1016/j.foodchem.2017.02.017

[94] Zijlstra DS, Korte J, Vries EPC, Hameleers L, Wilbers E, Jurak E, et al. High efficient semi-continuous extraction and in-line purification of high β-O-4 butanosolv lignin. Frontiers in Chemistry. 2021;**9**:655983. Available from: https://www.frontiersin.org/articles/10.3389/fchem.2021.655983/full

[95] Karlsson M, Vegunta VL, Deshpande R, Lawoko M. Protected lignin biorefining through cyclic extraction: Gaining fundamental insights into the tunable properties of lignin by chemometrics. Green Chemistry. 2022; **24**:1211-1223. Available from: https://pubs.rsc.org/en/content/articlelanding/2021/gc/d1gc04171a

[96] Zhang Z, Terrasson V, Guénin E. Lignin nanoparticles and their nanocomposites. Nanomaterials. 2021;**11**: 1336. DOI: 10.3390/nano11051336

[97] Lievonen M, Valle-Delgado JJ, Mattinen M, Hult E, Lintinen K, Kostiainen MA, et al. A simple process for lignin nanoparticle preparation. Green Chemistry. 2016;**18**(5):1416-1422. DOI: 10.1039/C5GC01436K

[98] Frangville C, Rutkevicius M, Richter AP, Velev OD, Stoyanov SD, Paunov VN. Fabrication of environmentally biodegradable lignin nanoparticles. ChemPhysChem. 2012;**13**: 4235-4243. DOI: 10.1002/cphc.201200537

[99] Yearla SR, Padmasree K. Preparation and characterisation of lignin nanoparticles: Evaluation of their potential as antioxidants and UV protectants. Journal of Experimental Nanoscience. 2016;**11**:289-302. DOI: 10.1080/17458080.2015.1055842

[100] Dai L, Li Y, Liu R, Si C, Ni Y. Green mussel-inspired lignin magnetic nanoparticles with high adsorptive capacity and environmental friendliness for chromium (III) removal. International Journal of Biological Macromolecules. 2019;**132**:478-486. DOI: 10.1016/j.ijbiomac.2019.03.222

[101] Chen L, Zhou X, Shi Y, Gao B, Wu J, Kirk TB, et al. Green synthesis of lignin nanoparticle in aqueous hydrotropic solution toward broadening the window for its processing and application. Chemical Engineering Journal. 2018;**346**:217-225. DOI: 10.1016/j.cej.2018.04.020

[102] Richter AP, Brown JS, Bharti B, Wang A, Gangwal S, Houck K, et al. An environmentally benign antimicrobial nanoparticle based on a silver-infused lignin core. Nature Nanotechnology. 2015;**10**:817-823. DOI: 10.1038/nnano.2015.141

[103] Lu Q, Zhu M, Zu Y, Liu W, Yang L, Zhang Y, et al. Comparative antioxidant activity of nanoscale lignin prepared by a supercritical antisolvent (SAS) process with non-nanoscale lignin. Food Chemistry. 2012;**135**:63-67. DOI: 10.1016/j.foodchem.2012.04.070

[104] Eerikäinen H, Watanabe W, Kauppinen EI, Ahonen PP. Aerosol flow reactor method for synthesis of drug

nanoparticles. European Journal of Pharmaceutics and Biopharmaceutics. 2003;**55**(3):357-360. DOI: 10.1016/S0939-6411(03)00005-5

[105] Ago M, Huan S, Borghei M, Raula J, Kauppinen EI, Rojas OJ. High-throughput synthesis of lignin particles (~30 nm to ~2 μm) via aerosol flow reactor: Size fractionation and utilization in Pickering emulsions. ACS Applied Materials & Interfaces. 2016;**8**(35):23302-23310. DOI: 10.1021/acsami.6b07900

[106] Gilca IA, Popa VI, Crestini C. Obtaining lignin nanoparticles by sonification. Ultrasonics Sonochemistry. 2015;**23**:369-375. DOI: 10.1016/j.ultsonch.2014.08.021

[107] Dallmeyer I, Ko F, Kadla JF. Electrospinning of technical lignins for the production of fibrous networks. Journal of Wood Chemistry and Technology. 2010;**30**(4):315-329. DOI: 10.1080/02773813.2010.527782

[108] Ten E, Ling C, Wang Y, Srivastava A, Dempere LA, Vermerris W. Lignin nanotubes as vehicles for gene delivery into human cells. Biomolecules. 2014;**15**:327-338. DOI: 10.1021/bm401555p

[109] Liu X, Yin H, Zhang Z, Diao B, Li J. Functionalization of lignin through ATRP grafting of poly(2-dimethylaminoethyl methacrylate) for gene delivery. Colloids Surfaces B: Biointerfaces. 2015;**125**:230-237. DOI: 10.1016/j.colsurfb.2014.11.018

[110] Qian Y, Zhang Q, Qiu X, Zhu S. CO_2-responsive diethylaminoethyl-modified lignin nanoparticles and their application as surfactants for CO2/N2-switchable Pickering emulsions. Green Chemistry. 2014;**16**:4963-4968. DOI: 10.1039/C4GC01242A

[111] Rangan A, Manchiganti MV, Thilaividankan RM, Kestur SG, Menon R. Novel method for the preparation of lignin-rich nanoparticles from lignocellulosic fibers. Industrial Crops and Products. 2017;**103**:152-160. DOI: 10.1016/j.indcrop.2017.03.037

[112] Melro E, Filipe A, Valente AJM, Antunes FE, Romano A, Norgren M, et al. Levulinic acid: A novel sustainable solvent for lignin dissolution. International Journal of Biological Macromolecules. 2020;**164**:3454-3461. DOI: 10.1016/j.ijbiomac.2020.08.128

[113] Sipponen MK, Lange H, Ago M, Crestini C. Understanding lignin aggregation processes. A case study: Budesonide entrapment and stimuli controlled release from lignin nanoparticles. ACS Sustainable Chemistry & Engineering. 2018;**6**:9342-9351. DOI: 10.1021/acssuschemeng.8b01652

[114] Kai D, Zhang K, Jiang L, Wong HZ, Li Z, Zhang Z, et al. Sustainable and antioxidant lignin-polyester copolymers and nanofibers for potential healthcare applications. ACS Sustainable Chemistry & Engineering. 2017;**5**:6016-6025. DOI: 10.1021/acssuschemeng.7b00850

[115] Chen N, Dempere LA, Tong Z. Synthesis of pH-responsive lignin-based nanocapsules for controlled release of hydrophobic molecules. ACS Sustainable Chemistry & Engineering. 2016;**4**:5204-5211. DOI: 10.1021/acssuschemeng.6b01209

[116] Farooq M, Zou T, Riviere G, Sipponen MH, Osterberg M. Strong, ductile, and waterproof cellulose nanofibril composite films with colloidal lignin particles. Biomacromolecules. 2019;**20**(2):693-704. DOI: 10.1021/acs.biomac.8b01364

[117] Liu C, Li Y, Hou Y. A simple environment-friendly process for preparing high-concentration alkali lignin nanospheres. European Polymer Journal. 2019;**112**:15-23. DOI: 10.1016/j.eurpolymj.2018.12.031

[118] Assis CA, Greca LG, Ago M, Balakshin MY, Jameel H, Gonzalez R, et al. Techno-economic assessment, scalability, and applications of aerosol lignin micro- and nanoparticles. ACS Sustainable Chemistry & Engineering. 2018;**6**:11853-11868. DOI: 10.1021/acssuschemeng.8b02151

[119] Gupta AK, Mohanty S, Nayak SK. Influence of addition of vapor grown carbon fibers on mechanical, thermal and biodegradation properties of lignin nanoparticle filled bio-poly(trimethylene terephthalate) hybrid nanocomposites. RSC Advances. 2015;**5**:56028-56036. DOI: 10.1039/C5RA07828H

[120] Qian Y, Qiu X, Zhong X, Zhang D, Deng Y, Yang D, et al. Lignin reverse micelles for UV-absorbing and high mechanical performance thermoplastics. Industrial & Engineering Chemistry Research. 2015;**54**(48):12025-12030. DOI: 10.1021/acs.iecr.5b03360

[121] Salas C, Ago M, Luvia LA, Rojas OJ. Synthesis of soy protein–lignin nanofibers by solution electrospinning. Reactive and Functional Polymers. 2014;**85**:221-227. DOI: 10.1016/j.reactfunctpolym.2014.09.022

[122] Qian Y, Deng Y, Qiu X, Li H, Yang D. Formation of uniform colloidal spheres from lignin, a renewable resource recovered from pulping spent liquor. Green Chemistry. 2014;**16**:2156-2163. DOI: 10.1039/C3GC42131G

[123] Yiamsawas D, Baier G, Thines E, Landfester K, Wurm FR. Biodegradable lignin nanocontainers. RSC Advances. 2014;**4**:11661-11663. DOI: 10.1039/C3RA47971D

[124] Li H, Deng Y, Liu B, Ren Y, Liang J, Qian Y, et al. Preparation of nanocapsules via the self-assembly of Kraft lignin: A totally green process with renewable resources. ACS Sustainable Chemistry & Engineering. 2016;**4**:1946-1953. DOI: 10.1021/acssuschemeng.5b01066